Storey's Guide to

RAISING
POULTRY

Storey's Guide to
RAISING POULTRY

Leonard S. Mercia

With revisions by Michael J. Darre, Ph.D., P.A.S.
Department of Animal Science
University of Connecticut
Storrs, Connecticut

STOREY
BOOKS

The mission of Storey Publishing is to serve our customers
by publishing practical information that encourages
personal independence in harmony with the environment.

Edited by Deborah Burns and Marie Salter

Copyedited by Arlene Bouras

Cover design by Renelle Moser

Cover images © John Colwell, Peter Dean, Grant Heilman from
  Grant Heilman Photography, Inc.

Back cover photography by PhotoDisc

Series design by Mark Tomasi

Production assistance by Deborah Daly and Susan Bernier

Line drawings by Elayne Sears, except for those on pages 11–13, 251, 255–258, and 293
  by Alison Kolesar; and on pages 14, 17, 25–27, 31 (top), 34, 36, 82, 88, 128, 130, 133,
  135, 153, 164–165, 175, 176, 211, and 289

Indexed by Susan Olason / Indexes & Knowledge Maps

Storey's Guide to Raising Poultry was previously published under the title *Raising Poultry the Modern Way*. This new edition has been expanded by 112 pages. Chapter 3, Poultry Biology, and Chapter 11, Game Birds and Other Poultry, were written by Michael J. Darre. All of the information in the previous edition has been reviewed and revised to represent the most comprehensive and up-to-date information available on the topic of raising poultry.

Printed in the United States by Versa Press
10 9 8 7 6 5

Library of Congress Cataloging-in-Publication Data

Mercia, Leonard S.
  Storey's guide to raising poultry / Leonard S. Mercia
    p. cm.
  Includes index.
  ISBN 1-58017-263-6 (alk paper)
  1. Poultry. I. Title: Guide to raising poultry. II. Title.
SF487 .M443 2000
  636.5—dc21   00-061245

CONTENTS

PREFACE

The incentive to write this book came in the early 1970s. It resulted from a sudden resurgence of requests for information relating to the management of small flocks of poultry. During the nearly 30 years the author was associated with poultry education through the Cooperative Extension Service, never did there exist as much interest in small flocks of poultry as during the 1980s.

For years we have witnessed the development of a highly specialized poultry industry made up primarily of large, automated production and marketing units. This evolution has taken place at the expense of the small poultry flock because of diminishing profits and the difficulties facing small-flock owners trying to market their products. Suddenly, there developed a renewed interest in raising a few chickens, turkeys, waterfowl, or other types of poultry. Why this sudden turnabout?

The desire to produce fresh eggs and poultry for the family probably ranks high among the reasons for starting a small poultry flock. However, a few birds may also provide family members with an excellent chore responsibility, a hobby, or even a source of limited income. Whatever the reason, success in such a venture requires a knowledge of poultry husbandry and many other related subjects.

The need exists for a handy reference manual with complete and practical information for the small poultryman or the prospective poultryman. This book is written with that in mind. It treats practically all phases of poultry production and processing, and includes sections on chickens, turkeys, waterfowl, game birds, and other poultry.

The first chapter attempts to help answer the question "Should I raise a poultry flock?" and includes a discussion of the possible types of poultry projects and some of the advantages and disadvantages of each. The chapter on the laying flock will probably interest the largest number of readers, since this seems to be the project that appeals to most people. Much of the chapter centers on questions frequently asked of the author. It therefore covers topics such as types and sources of stock, management and feeding practices, housing, and equipment needs. It contains plans for small poultry houses, poultry furniture, a small incubator, and other equipment. One chapter is devoted to meat-bird production. Other chapters treat turkey, waterfowl, and game-bird production. You will find information on such skills as culling, beak trimming, and caponizing; on recognition of diseases and their treatment; and on egg care, incubation, and egg and poultry processing. To enhance the value of the book as a reference, it includes lists of additional reading materials, poultry diagnostic laboratories, cooperative extension system offices by state, associations, sources of supplies and equipment, and a glossary.

Because the book includes information on more than one species, and many of the subjects discussed require references to other subjects, some repetition was unavoidable. Then, too, it was felt that repetition might be helpful to those readers who have an interest in only one section of the book or who are able to read only one section at a time. Cross-references are used to avoid duplication as much as possible.

Although intended primarily as a guide and reference for prospective poultrymen and small-flock owners, *Storey's Guide to Raising Poultry* also has value as a text or a supplement for practical poultry production courses or science projects in the schools. The new chapter on poultry biology may be of special interest to students.

ONE

SHOULD I RAISE POULTRY?

The care of a small poultry flock can be a rewarding experience as a hobby, for show, to help fill the family refrigerator or freezer, or to provide eggs and poultry for sale. Poultry are relatively easy and inexpensive to raise, take little space, and are fun to watch. With good husbandry and flock management, the poultry project can be a source of efficiently produced food. To lower the costs of feed, commercially produced poultry feeds or well-formulated, home-mixed diets can be supplemented to some extent with scraps from the table, some green garden crops, or fine lawn clippings. Selecting the right stock for your particular type of project — whether it be ornamental birds, laying birds, broilers, roasters, capons, turkeys, waterfowl, or others — will optimize results. Moreover, surplus eggs and poultry may be sold at a profit.

Benefits of Raising Poultry

Poultry have become very popular as 4-H and school projects because of the relative ease and low cost of rearing them. The poultry project becomes a way for youth to learn to care for living things as well as to manage their time and money. Poultry projects such as growing broilers or layers can also supplement the family's food supply.

We are currently experiencing an increased demand for chicks and breeding stock, due to a resurgence of interest in the production of food at home. Often the responsibility for the care of the poultry project is given to the younger members of the family. This can be a well-rounding experience for young people, fostering a basic knowledge of such things as egg care, incubation, feeding and management of growing birds, and, if properly undertaken, the development of such important skills as record keeping and money management.

A valuable by-product of the poultry project that is frequently overlooked is the manure. Poultry manure is high in fertilizer value and is excellent for the garden and a valuable addition to the compost pile.

Commercial versus Small Enterprises

The modern commercial poultry industry is perhaps best characterized as being a highly specialized, highly commercialized agricultural industry. It is no longer a small business on many farms, but a large business on fewer and fewer farms. The keys to success in the commercial poultry business are frequently size, volume, and efficiency. The necessary capital investment is high; there are some pitfalls in the business, and a profitable market for large-volume production is nearly essential if the enterprise is to be successful.

Small poultry enterprises with a few hundred or a few thousand birds can be profitable if the products can be marketed at a substantial margin directly to stores, restaurants, hospitals, institutional users, or individual consumers. Even the small family flock can return a profit when products are marketed directly to the consumer at a substantial markup. These outlets, however, are not always free from problems. Competition for markets is keen, and profit margins sometimes erode. Moreover, there is the problem of providing a volume of sizes, quality, and service that is often difficult for the small producer who sells to wholesale outlets.

Consider Market Potential, Zoning Restrictions, and Your Commitment

If you plan to sell your products and make a profit, you need to study carefully the market potential in your area. Without profitable markets, any poultry enterprise is almost certainly doomed to failure.

Before starting a poultry flock, check the local zoning ordinances. Animal projects are not permitted in certain residential areas. Poultry flocks can bring about such problems as odors, flies, fleas, rats, and mice. There is also the possibility of complaints about noise, especially if there are adult males in the chicken flock. Nothing manages to anger neighbors more than the early-morning crowing of a rooster. Good management will prevent many of these nuisances, but it is best to research thoroughly all possible problems before embarking upon your own poultry project.

Should you decide to raise poultry, it is, in most instances, a 7-day-a-week job. There are the daily management chores to be done — feeding, watering, collecting eggs, and checking the birds. And last, but not least, there is the needed investment in poultry housing and equipment, as well as the cost of feed, birds, and supplies. Consider these factors carefully before you begin a poultry project.

Types of Poultry Projects

There are many types of poultry projects that are of potential interest to poultrymen and students alike. It is for space reasons alone that we will limit our discussion to the five types that are of interest to the greatest number — the laying flock, meat chickens, turkeys, waterfowl, and game birds. A brief discussion of other poultry projects is also included

The Laying Flock

The laying flock is the most popular type of poultry project. Good stock properly managed will lay eggs throughout the year. Most laying birds start laying at 18 to 24 weeks of age. The length of their laying cycle is normally 12 to 14 months. Any stock that is bred for high egg production is suitable for the laying flock; however, many small-flock owners rear a dual-purpose bird, such as Rhode Island Reds, Barred Plymouth Rocks, New Hampshire Reds, or sex-link crosses, because they get good egg production along with a meatier carcass at the end of lay. A good laying bird, under proper management conditions, should lay about 20 dozen eggs during a laying cycle. Then, too, there is the possibility of an occasional meal of fowl when poor producers are culled from the flock.

Meat Chickens

Specialty meat birds such as capons and roasters are particularly suited to small-flock enterprises. Broilers, as a general rule, cannot be grown on the small farm as economically as they can be grown commercially.

However, some individuals still choose to grow their own broilers to control the type of feed and feed ingredients the birds consume, and to have available their own freshly slaughtered birds to eat. Broilers are young chickens used for frying, broiling, or roasting, and are usually 6 to 8 weeks of age when processed. Commercial broilers are selectively bred to produce meat efficiently. In the early days of the development of the broiler industry, broilers were the by-product male birds from straight-run flocks being grown primarily for replacement pullets for laying flocks. At roughly 13 weeks of age, the males were removed from the flock and slaughtered as broilers. Very few broilers are now produced as a by-product of egg production.

Roasters are 8- to 14-week-old chickens weighing 5 pounds (2.3 kg) or more. Capons are castrated males grown for 12 to 18 weeks to weights of 8 pounds (3.6 kg) or more. The costs of raising these chickens to heavier weights are considerably higher than the costs of raising broilers. These tender-meated, well-finished birds are, however, sought after and demand a much higher price per pound in the market.

Turkeys

Turkey production makes an excellent project for fun or profit. Although it was once thought to be hazardous and difficult to raise turkeys, this is no longer true, thanks to the available modern-day drugs and management know-how. Still, certain precautions have to be observed. Blackhead disease is common among turkeys, and they should be reared separately from chickens to prevent problems with this disease (see page 271 for more on blackhead). It is possible to raise turkeys either in confinement or on range.

A turkey project can be excellent for anyone, young or old. It can provide a valuable responsibility and learning experience for young people, and with good stock, good management, and good marketing, a profit can be realized. Turkeys can be raised successfully in virtually any climate if they receive the proper management and nutrition and are protected against diseases, predatory animals, and exposure to extreme weather conditions.

Turkey production for roasters has the added advantage of being a relatively short-term project, since it requires only about 4 to 6 months to prepare for the poults, grow them, process and market them, and clean the facilities in preparation for the next flock. Turkey broilers require only 10 to 14 weeks to finish.

Fresh native turkey, properly dressed, is hard to beat for juiciness and flavor, so it is a natural choice for the production of food for home consumption.

Waterfowl

As in other areas of the poultry industry, some farms specialize in the commercial production of ducks and geese. Many small flocks are also raised to occupy the farm pond or as a hobby or sideline to other farm enterprises. Ducks and geese make an excellent hobby, and many are raised for exhibition purposes or multiplied for sale as ducklings or goslings.

Well-finished ducks and geese are a nutritious and tasty food. Some breeds of ducks, surprisingly enough, surpass some chickens in egg-laying ability.

Rearing waterfowl is relatively easy. The young are quite hardy, and by following reasonable management and feeding programs, such a project can be successful and profitable. Ducks and especially geese are excellent foragers and can pick up a great deal of their nutritive requirements during the warm-weather months in the form of green feed, insects, and worms. In confinement they can cause damp litter conditions unless special waterers and management procedures are used.

Game Birds

Game birds are grown for a number of reasons. They provide some growers with an excellent income when produced and marketed in an area where there is a good demand. Included in this group are pheasants, partridge, and quail. Breeders can sell meat birds or breeding stock and eggs. There is a demand in some areas for birds to restock hunting farms and preserves.

Other Poultry Projects

The American Standard of Perfection, which is published by the American Poultry Association, recognizes more than three hundred breeds and varieties of domesticated chickens, including bantam fowl, turkeys, and waterfowl. The standard describes birds on the basis of class, breed, and variety.

The term *class* is usually used to designate groups of standard breeds developed in various regions; thus, the class names — for example, American, English, Mediterranean. The main breed differences are those of body shape and size. Feather color and color pattern, as well as comb type, distinguish the different varieties.

There are other characteristics common to most breeds within the classes. For example, American breeds commonly lay brown-shelled eggs, and those breeds within the Mediterranean class lay white-shelled eggs.

There is quite a large number of people with an interest in standard-bred poultry. The interest of these "poultry fanciers" is in breeding and multiplying birds for exhibition at fairs and poultry shows. In addition to competing for prize money and trophies, they usually sell breeding stock and eggs. It provides some income and is also an interesting hobby. This project also provides eggs and poultry for the table.

Pigeons

Some bird fanciers breed and exhibit pigeons and enter them in shows. They have the opportunity to sell breeding stock, as well.

Other pigeon breeders raise strictly for meat production. Young squabs are sold in specialized markets for a good price — they are a real delicacy and may provide a substantial income if grown on a commercial scale.

A number of pigeon breeders are interested in producing racing pigeons. They usually participate in long-distance races with the hope of winning prize money.

Guinea Fowl

Guinea fowl is a special breed grown for the production of meat. It provides all dark meat and brings a good price in specialized markets. Guinea fowl is also exhibited at poultry shows and, like other exhibition fowl, may provide the fancier with a profit from both the sale of breeding stock and hatching eggs. It is also gaining popularity as a way to control the insect population around the home and to reduce the deer-tick population and, hence, the incidence of Lyme disease.

Exhibition Fowl

The management and housing principles for exhibition fowl are much the same as those for the production of small flocks; however, breeding and selection of birds for exhibition require a knowledge of breed and variety characteristics, of genetics, and of fitting and training the birds to be shown.

RAISING POULTRY FOR SHOW

Jim is a 4-H member who became interested in raising poultry after seeing several breeds of show birds at the county fair. He started with a few Buff Cochin bantams he bought from one of the exhibitors. He then learned that the 4-H club in his area had a few members with poultry and so he joined the club. He learned proper management and health care of his birds and how to prepare them for show. He has a 15-foot by 20-foot (4.6–6.1 m) building with smaller breeding pens. He keeps records on feed use and other expenses and uses the money he gets from show winnings and the sale of fertile eggs to help finance some of those costs.

The club participates in 4-H poultry showmanship contests, and Jim has become an experienced showman. He now breeds and shows champion Cochins, Japanese, and Silkies. Other members of his 4-H club raise chickens for egg production, one family raises broilers each year, and all the club members help process the birds. When asked why they raise chickens, members say that chickens are fun to watch, and they like the fresh eggs and meat.

TWO

POULTRY HOUSING AND EQUIPMENT

The type of poultry house you'll need will vary with the type of project and how you plan to get started. For example, baby chicks and turkey poults require tighter, more comfortable housing than do older birds. If you plan to begin the project with started birds, the house may not have to be as tight or as well insulated. In warm climates, the house may have open sides covered with poultry netting. It may be equipped with adjustable awnings for use during inclement weather. Waterfowl require very little in the way of housing as adults; however, young ducklings and goslings, though relatively hardy, will need good housing during the early brooding period.

The type of house will also depend, to a certain extent, upon the management system you use. Some differences in design and size may be needed to accommodate different types of equipment and furniture. If cages are utilized for laying flocks, the building construction and design must be considered to ensure that the equipment fits in the building. Material-handling chores such as feeding, cleaning, watering, and gathering eggs should also be considered when planning the poultry house.

The poultry house need not be elaborate or expensive. Frequently, an old building can be remodeled to accommodate a small flock of poultry. Sometimes a pen built within a large existing structure will function quite well for small flocks. Remodeling of old buildings for large production units is seldom advisable; although some old structures are suitable for the production of broilers, pullets, or turkeys, most are not readily adapted to the efficient handling of birds. Material-handling chores for young growing birds are not so critical, but it is an important consideration for large laying projects.

There are basically two types of poultry housing. One type is designed for floor management systems, the other for cage systems.

Floor Housing

Floor housing, also called litter or loose housing, allows the birds free access to the poultry house or pens within the house. The advantage of the floor system is that it permits flexibility in the use of the house. It can be adapted for brooding and growing or managing most any type of poultry flock, including broilers, roasters, capons, layers, turkeys, or waterfowl. This system is well suited for the small farm or family flock, and most of our consideration will be given to this type of housing.

Cage Housing

A word about cages: This is the system most frequently used by commercial egg producers today. There are several reasons for its popularity. Floor-space requirement per bird is less, thus reducing per-bird cost of housing. Cage systems are readily adapted to labor savings in materials handling. The use of automatic feeders and waterers, the belt collection of eggs, and manure-handling devices are naturals with cages.

Another important advantage of cage units is that eggs do not have to be collected as frequently because of the egg rollaway feature. Cage housing and management systems permit the care of more birds per person than does the floor system. For most, cage facilities offer better control of parasites and eliminate many of the troublesome litterborne diseases sometimes found with floor birds. Most breeding farms still use the floor systems of housing for the mating of breeders and the production of fertile eggs. Breeding cages are available, and can be used successfully for some strains of light breeders. Cages can also be used for brooding and rearing of chicks in controlled-environment housing, reducing the need for litter as well as the spread of disease between chicks.

The Ideal Poultry House

The ideal poultry house, or poultry pen, meets the specific needs of its occupants. It must provide a clean, dry, draft-free, comfortable environment for the birds throughout the year. If you are to brood young birds in the house, you must build it well enough to permit a comfortable environment and efficient use of fuel. It must be tight enough to prevent drafts on the birds and help maintain a uniform temperature economically.

Poultry facilities should have floors that can easily be cleaned and disinfected. In many areas, concrete floors are used because they are easy to maintain and keep pests from burrowing through the floor. Good-quality CDX plywood coated with epoxy or another sealant also works well.

Most poultry pens or buildings should provide for ventilation to control moisture in the pens, remove gases, and conserve or dissipate heat. In the northern climates, the houses should have insulation in both the sidewalls and the ceilings. Piped-in water is desirable but may create problems in severely cold climates, especially in the small-flock situations where body heat is not sufficient to keep the pen temperatures above freezing. It is possible to correct this problem through the use of electric heat tapes and water warmers. Houses should have artificial light to provide the right lighting program for both layers and growing stock. Finally, the poultry house should be strong enough to withstand high winds and snow loads, if necessary.

The house plan shown is for a small flock and provides room for storage of feed and supplies.

Poultry House Plans

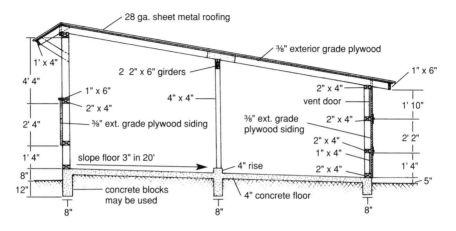

Cross section

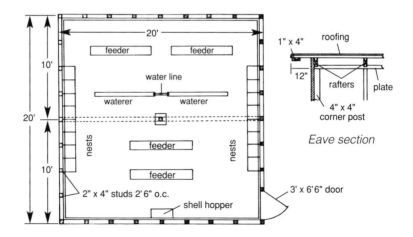

Plan

Eave section

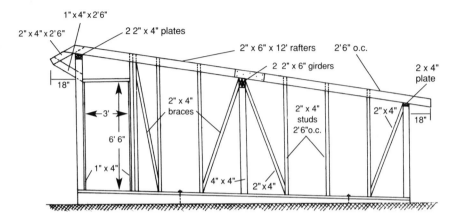

Side framing

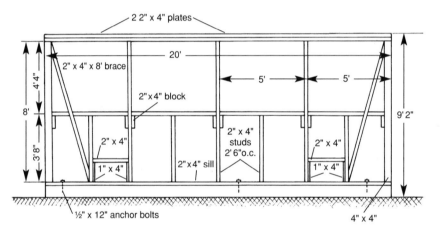

Front framing

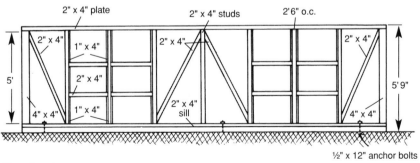

Rear framing

1" mesh poultry netting — use plastic
on outside during cold weather

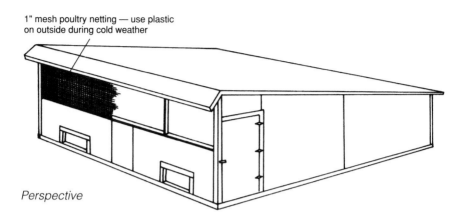

Perspective

Selecting a Site

When selecting a site for a poultry house, consider the soil drainage, air movement, location of the dwelling, and water supply. Remember, there is the possibility of odors, flies, fleas, rats, and mice. Good soil drainage ensures dry floors, which will help prevent wet litter, dirty eggs, disease, and other problems.

Locate the poultry house where the prevailing summer winds will not carry odors to the dwelling. The site should be large enough to provide for expansion, if this should seem advisable in the future. Also consider the location of access roads, electric lines, future buildings, and yards.

A site on relatively high ground with a south or southeast slope and good natural drainage is desirable. Siting a poultry house at the foot of a slope where soil or air drainage is poor, or where seepage occurs, is unwise.

If you locate the house on a hillside, make sure the site is graded so as to carry surface water away from the building. In some instances, tile drains will be necessary to adequately carry the water away from the building's foundation.

In areas of severe winter weather, place most of the windows in the front of the building and face the house south to take advantage of the sunlight and solar heating. The pen will be warmer, the litter drier, and the birds more comfortable. The closed or back side of the house should provide maximum protection against the northwest wind. Where the prevailing winter storms are from the west, however, orient the house facing east. Wider houses are often placed with their long dimensions north and south to provide lighting on both sides. If the house is oriented with the front window area to the south, solar heat can be controlled by the width of roof overhang to provide shade in the summer months when the sun is high on the horizon and to permit sunlight in the pens in the winter when the sun is low.

Portable Housing

If you have just a few birds, and want to provide them with access to clean, fresh grass or pasture throughout the summer, but within a fenced-in environment, then build a small house on skids or runners. This will allow the movement of the house from one place to another with relative ease. Skids or runners can be made of 4 x 4 or 6 x 6 pressure-treated lumber, with one or both ends cut at an angle or rounded, so it will not dig into the ground when pushed or pulled. By connecting the two runners in an H pattern with boards about one-eighth of the way in from each end, you will stabilize them and prevent them from turning inward as you move the house. You can drill a hole through one of the crossbeams and place an eye-ring and bolt through it to secure a rope or chain when towing. Use movable fencing, and every few weeks you can move the whole setup to a fresh place on your property.

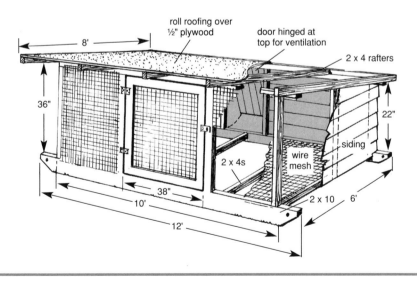

Space Requirements

The size of the poultry house will depend upon the type and number of birds to be housed, as well as the management system to be used. Various age groups and species of birds require different amounts of floor space for optimum results. If the house is designed as a laying house, and a cage management system is to be used, the size of the building will depend upon the dimensions of the cages, the type of cage, and the number of birds to be housed per cage. In addition to space needed for the birds, there should always be additional space for storage of feed, supplies, and equipment.

Floor-space requirements vary according to the type of project, the size or age of the birds, and the management system in use (see chart). It is important to have adequate floor space to prevent such problems as cannibalism, poor growth or poor egg production, and morbidity or mortality.

Floor Space Requirements for Poultry

TYPE OF BIRD	AGE OF BIRD	FLOOR SPACE* (SQ. FT.)
Chicks	0–10 weeks	0.8–1.0
	10–maturity	1.5–2.0
Layers	Brown egg	2.0–2.5
	White egg	1.5–2.0
	Meat breeders	2.5–3.0
Broilers	0–8 weeks	0.8–1.0
Roasters	0–8 weeks	0.8–1.0
	8–12 weeks	1.0–2.0
	12–20 weeks	2.0–3.0
Turkeys	0–8 weeks	1.0–1.5
	8–12 weeks	1.5–2.0
	12–16 weeks	2.0–2.5
	16–20 weeks	2.5–3.0
	20–26 weeks	3.0–4.0
	Breeders (heavy)	6.0–8.0
	Breeders (light)	5.0–6.0
Ducks	0–7 weeks	0.5–1.0
	7 weeks–maturity	2.5
	Breeders (confinement)	6.0
	Breeders (yarded)	3.0
Geese	0–1 week	0.5–1.0
	1–2 weeks	1.0–1.5
	2–4 weeks	1.5–2.0
	Breeders (yarded)	5.0

*Floor space needs vary with type and age of bird and type of management system.

Doghouses Are for the Birds

Some people who have just a few birds for eggs or for ornamental purposes and need shelter only at night for their birds use a modified doghouse design. A doghouse with dimensions of about 3 feet by 5 feet (0.9 x 1.5 m) or larger is ideal for housing a few birds. Connect the roof to one of the walls with hinges, which allows the roof to lift open and gives you easy access to the inside. Cut ventilation slots or holes, cover them with hardware wire, and put in a locking door. This simple doghouse design provides a secure space for your birds.

Insulation

Removal of moisture from the building is one of the main problems of poultry housing. The average moisture content of freshly voided poultry manure is 70 to 75 percent. The moisture content of the manure varies with the type of feed, feed ingredients, temperature in the building, and even the type of bird. Waterfowl, for example, tend to produce wetter droppings than do chickens and turkeys. Also, respired moisture and the moisture of the air brought in by the ventilation system must be removed from the building.

Insulation will provide optimum bird comfort and prevent excessive moisture problems if the heat is adequate. In most areas of the United States, the poultry house should be insulated. The purpose of insulation is to conserve heat in the cold climates and to keep out the heat in warm climates. It should be pointed out that light-colored reflective roof and wall surfaces are helpful in keeping the pens cooler in warm climates. Because warm air carries more moisture than cold air, insulation serves to conserve heat in cool climates not only to provide birds warmth but also to permit removal of moisture from the building. It is important to keep the litter dry and to prevent buildup of ammonia in the house. Excessive ammonia can be detrimental to the birds and uncomfortable for the operator.

It was once thought that only housing in the northern climates should be insulated. However, the development of the so-called controlled-environment house for high-density operations has changed this philosophy, and we now find more of this type of housing being built in the South as well as in the northern climates, particularly for cage management systems. Controlled-environment houses are windowless structures with artificial light and fan ventilation.

Strategies for Small Flocks in Northern Climates

In the northern part of the United States, the small flock (fewer than several hundred birds) is not without its special problems. Even the well-insulated poultry house may not provide optimum conditions during severe winter weather unless artificially heated. It is essentially impossible or highly impracticable to insulate buildings well enough to conserve the heat given off by a small number of birds. Frozen waterers, frozen pipes, or even frozen combs may result, causing a drop in egg production. Cold buildings also cause excessive feed consumption. Well-insulated pens within large existing buildings may be one answer to some of the above-mentioned problems, especially if other animals are housed in the building and help provide some of the warmth within the structure. Houses that are well protected from the wind are also much easier to keep warm and comfortable in winter weather.

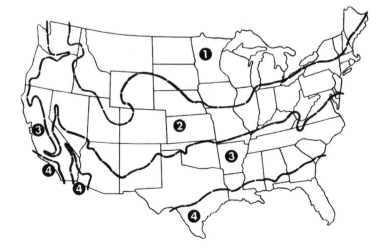

Climate influences poultry house design, especially for the small flock. This farm building zone map shows the four basic climatic zones in the United States, based on January temperatures and relative humidity.

The chart below gives the recommended insulation values for poultry houses in the different climatic zones. The resistance of insulation values of various building materials is shown opposite.

Ventilation

Ventilation provides bird comfort by removing moisture, ammonia, and other gases from the building; provides an exchange of air; and controls the environmental temperature in the pen.

There are two types of ventilation systems commonly used in poultry houses: the natural or gravity system and the forced-air system.

Gravity System

The gravity system may use windows, special slot inlets, or flues to provide air movement. The incoming cool air replaces the warmed, moisture-laden air going up through flues in the ceiling or through the top of opened windows. If flues are used in cold climates, they must be insulated to prevent condensation of moisture that drips back into the pen. Flues must also extend well above the roof to provide the necessary draft to remove the warm, moisture-laden air. This approach to ventilation works satisfactorily in an insulated house where inlets or windows are properly adjusted. The gravity method of ventilation has been replaced

Recommended Minimum Insulation Values for Poultry Houses by Zones

LOCATION	R-VALUE	
	WALLS	CEILINGS
Zone 1:		
Colder parts	8–10	10–15
Warmer parts	6	12
Zone 2	5	10–12
Zones 3 and 4	2	5

R-value — thermal resistance, or insulation value — is defined as the number of degrees of difference in temperature between the inside and outside surfaces of a wall that will permit 1 British thermal unit (BTU) of heat to pass through 1 square foot (0.09 sq m) of the wall per hour. Source: U.S. Department of Agriculture.

gradually as the size of poultry operations has increased. Opening and closing windows with changes in temperature and wind conditions is a very time-consuming chore. Maintenance of the windows is also time-consuming and costly.

R-Values of Various Building Materials

MATERIAL	THICKNESS IN INCHES	R-VALUE
Air space	3/4–4	0.91
Blanket insulation (glass or rock wool)	1	3.70
Blanket insulation (glass or rock wool)	2½	9.25
Blanket insulation (glass or rock wool)	3	11.10
Building paper		Negligible
Cinder block	8	1.73
Concrete	10	0.80
Concrete block	8	1.11
Fill insulation:		
Shavings	3⅝	8.85
Shavings	5⅝	13.70
Sawdust	3⅝	8.85
Sawdust	5⅝	13.70
Fluffy glass fiber, mineral, or rock	3⅝	13.40
Homosote	½	1.22
Insulation board (typical fiber)	½	1.52
Insulation board (typical fiber)	25/32	2.37
Plywood	3/8	0.47
Sheathing and flooring (softwood)	3/4	0.92
Shingles (asphalt)		0.15
Shingles (wood)		0.78
Siding (drop)	3/4	0.94
Siding (lap)		0.78
Surface (inside)		0.61
Surface (outside)		0.17
Window (single glass)		0.10
Window (double glass in single frame)		1.44

Forced-Air System

Forced-air ventilation gives the best control of air movement. Fans are used to move out the warm, moisture-laden air and ammonia, and bring in cool, fresh air for bird comfort through adjustable slots located in the walls near the ceiling. Fan capacity is determined by the type, age, and number of birds in the house. Generally, a capacity of 3 to 5 cubic feet (0.08–0.14 cu m) feet per minute (cfm) per bird is recommended. For example, if you are building a house for two thousand growing or laying birds, design it with a fan capacity ranging from a low of 500 to a high of 10,000 cfm (14.2–283 cu m). You can accomplish this flexibility of air movement with multiple fans, variable-speed fans, or a combination of both. It permits a small amount of air movement for young chicks or for cold weather conditions, and a larger volume of air movement for older birds or warm weather conditions.

In cold climates, 5 cfm (0.14 cu m) per bird would be too much air movement for bird comfort, so the ventilation rate is cut back by the use of thermostats or timers that control the fans. The optimum temperature in cold climates for all but young birds under 6 to 7 weeks of age is approximately 55°F (13°C). Do not allow fans to cycle on and off at frequent intervals. Fan cycling occurs when pen temperatures are cooled down too quickly by overventilating. When that occurs, the system is unable to remove the maximum amount of moisture from the pens, and the birds are exposed to uncomfortable temperature conditions. You can prevent fan cycling with proper thermostat settings, and by restricting the air inlets in cold weather, to maintain a stable relationship among pen temperature, incoming air, and exhausted air. In the warm months, open the inlets wide to permit a maximum flow of air for removal of moisture and ammonia and for cooling the birds. Usually, you can satisfactorily ventilate houses for small flocks through the windows or slot inlets.

Lighting

An essential component of the birds' environment is light. Light is an important factor in the maturation of laying and breeding birds. It can control when a bird comes into production and when it goes into a molt. It can keep a layer producing eggs for an extended period of time.

Light can be supplied naturally through windows or artificially in environmentally controlled buildings, or by supplementing natural light with artificial light. It has three important components that must be considered: wavelength, intensity, and duration.

Wavelength

Wavelength determines the color of the light. Visible light is produced in the range of about 400 nanometers (nm) to 700 nm, or from near ultraviolet to near infrared. Humans have a peak sensitivity of about 555 nm, in the yellow-green portion of the spectrum. Chickens and other poultry, on the other hand, peak at about 580 nm, more in the orange-red portion. It is for this reason that lamps producing light in the orange-red portion of the visible spectrum are preferred for rearing poultry. Incandescent lamps produce a large portion of their light in the red and infrared range, but are not efficient, because half or more of the electrical energy they use is converted to infrared heat energy, not light. The newer compact fluorescent (CF) lamps in the 2500- to 3200-degree kelvin (K) range (warm white) produce less infrared, are more efficient, and are good for use with poultry.

Intensity

Intensity is the brightness of the visible light and is measured in foot-candles (FC), or lux. About 10.76 lux equals 1 FC. For most poultry facilities, the ability to provide light intensities ranging from 0.5 FC to 10 FC at bird level is adequate. You can do this by changing the size of the lamp or with the use of dimmers. Different types and ages of birds require different levels of light intensity. (See discussions of specific species.)

Duration

Duration or day length is important for maturation and egg production. Generally, more than 11 hours of light per day is photostimulatory and may cause young birds to mature sooner and come into egg production. Increasing hours of light simulates springtime and causes maturation of the reproductive tract, whereas decreasing hours of light will cause a bird in production to molt (shed old feathers and grow new ones). Therefore, a general rule of thumb is never to expose adolescent growing birds to increasing hours of light unless you are ready to stimulate egg production, and never to expose laying hens in production to decreasing hours of light unless you want to molt them. Laying hens are typically maintained on 15 to 16 hours of daylight, to simulate the longest natural day length.

Equipment Requirements (per 100 Birds)

BIRD TYPE	AGE (WEEKS)	BROODERS		FEEDER	WATERER
		HOVER TYPE	INFRARED LAMP		
Chicks (broilers, roasters, capons, pullets)	0–2	700 sq. in.	2	1 feeder lid to 7 days 100" of trough or 3 hanging	20" trough or 2 1-gal.
	2–6	700 sq. in.	2	200" of trough or 3 hanging	40" trough or 3 1-gal.
	6–maturity	—	—	300" of trough or 4 hanging	96" of trough or 1 automatic
Layers	22–market	—	—	300" of trough or 4 hanging	96" trough or 1 automatic
Turkeys	0–2	1200 sq. in.	2–3	1 feeder lid to 7 days 200" trough or 3 hanging	20" trough or 2 1-gal.
	2–6	1200 sq. in.	2–3	400" trough or 5 hanging	96" trough or 1 automatic
	6–market	—	—	480" trough or 6 hanging	120" trough or 1 automatic
Ducks and geese	0–2	1200 sq. in.	2–3	1 feeder lid to 7 days 100" of trough or 3 hanging	20" trough or 2 1-gal.
	2–4	1200 sq. in.	2–3	200" of trough or 5 hanging	50" trough or 3 automatic
	4–market	—	—	200" of trough or 5 hanging	50" trough or 3 automatic

Note: *A trough feeder or waterer 2 feet (61 cm) long and open on both sides provides 48 inches (121.9 cm) of space.*

Management Systems

There are essentially three types of management systems used for the care of poultry. The first utilizes the housing floor; the second, slats or wire mesh; and in the third the poultry are housed in cages.

Floor Management System

The management system most widely used, until a few years ago, was the floor or litter management system. This is still the most popular method for small flocks, breeder flocks, growing birds, broilers, roasters, turkeys, and waterfowl. It permits the birds freedom in the entire pen or building. Litter materials such as wood shavings, ground corncobs, sawdust, sugar-cane, rice hulls, and finely chopped straw are used on the floor to insulate it for bird comfort and to absorb moisture. Litter also helps control disease and enhances egg cleanliness in laying flocks. In addition to the necessary feeding and watering equipment, the system may use nests or roosts or dropping pits, depending upon the type of bird housed.

A variation of this confinement management system uses a fenced yard adjoining the house or pen. Yards are useful for small flocks, particularly where floor space is limited. Grass sod in the yard can also provide some forage for the birds. Use of the yard depends upon the age and type of bird housed, climate, and other factors.

Slats or Wire Mesh

Another type of management system uses slats or wire mesh as the floor. This system has been used for laying birds, turkeys, and waterfowl. The entire floor is slats or wire. No litter is used and roosts are not required. The birds' droppings fall through the wire or slats into a pit and are generally cleaned out periodically or when the birds are removed.

A variation of this method incorporates either slats or wire and a litter section. The feeders and waterers are usually placed over the wire or slat sections, and the fecal material drops into a pit, either shallow or deep. The pit and the slatted or wire section are usually located on the sides of the pen with a litter section in the center. This system has several advantages, in that the birds spend approximately 75 percent of their time on the slats or wire sections, eating and drinking. Therefore, the majority of the droppings go into the pits, helping keep the litter section dry. This greatly simplifies the job of ventilating the building and results in cleaner litter and cleaner nests and eggs. The same types of feeders, waterers, and nests are used for litter-floor houses as for the slat- or wire-floor houses.

Cages

The third type of management system uses cages. As mentioned earlier, cages have become the most popular means of managing market egg birds; most of the large commercial laying houses are now equipped with wire cages. Cages have sloped floors so eggs roll to the front for easy collection. The eggs need not be collected as frequently, since the birds can't set on them. Cages are also coming into use to brood pullets to be housed in laying cages. Thus far, attempts to raise meat birds in cages haven't been particularly successful.

Furnishing the Poultry House

Furnishings for the poultry house depend upon the type and size of the project. If it is a growing project, the major requirements will be brooding equipment, feeders, and waterers. A laying-bird project will require feeders, waterers, nests, roosts, and a broody coop for birds that go broody; that is, for those that are ready to brood, or sit on eggs. If laying cages are used, no nests, roosts, or broody coops are needed.

The equipment need not be fancy. Feeders and waterers should be designed to service the birds efficiently with a minimum of waste or spillage. There must be enough equipment to give each bird an equal chance to use it. The amount of equipment required varies with the type of bird, age, and size.

Feeders

Feeders should be large enough to supply the flock's needs for a day or more without wasting feed. Proper feeder design is an important consideration in avoiding feed waste. The use of an antiroost device, such as a reel or a spring-loaded wire on top of the feeders, will help prevent waste. A lip on the side of the feed trough will prevent birds from throwing out feed. Feed troughs should never be filled more than one-third to one-half full to avoid feed waste. The size, height, and construction of the feeder are very important if feed waste is to be kept to a minimum. To minimize feed waste, make the troughs so that you can adjust the height as the birds grow. The alternative is to change to a larger feeder as the birds become larger. Less feed is wasted if the lip of the feeder is level with the top of the bird's back as it stands. Some of the feeders commonly used and wooden feeders you can build are illustrated next.

Feeders

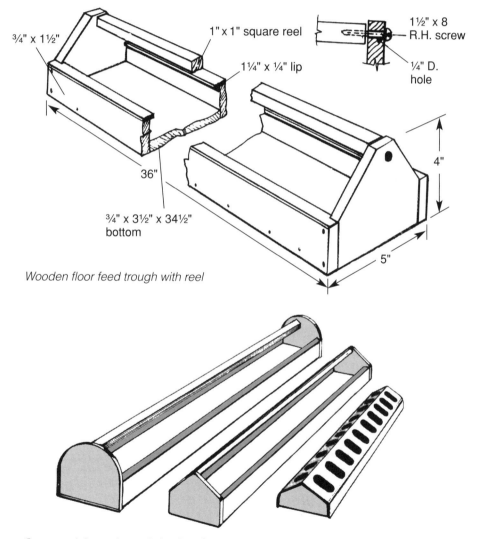

Wooden floor feed trough with reel

Commercial metal trough feeders for various age groups

Feeders (continued)

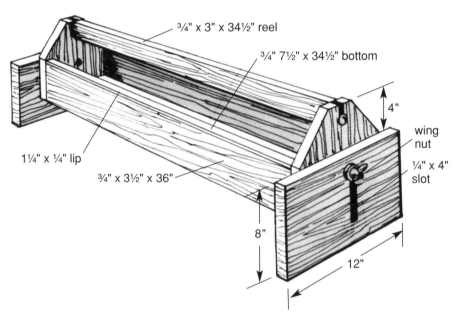

¾" x 3" x 34½" reel

¾" 7½" x 34½" bottom

4"

wing
nut

¼" x 4"
slot

1¼" x ¼" lip

¾" x 3½" x 36"

8"

12"

Adjustable height floor feed trough with reel

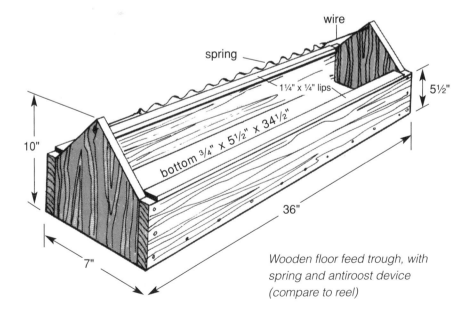

wire

spring

1¼" x ¼" lips

5½"

bottom ¾" x 5½" x 34½"

10"

36"

7"

*Wooden floor feed trough, with
spring and antiroost device
(compare to reel)*

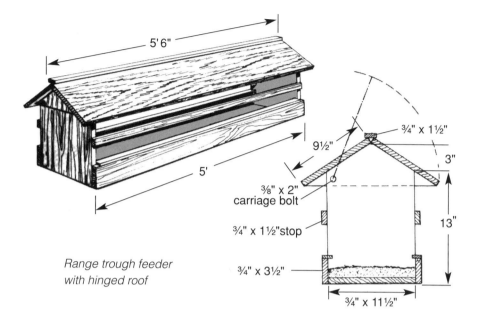

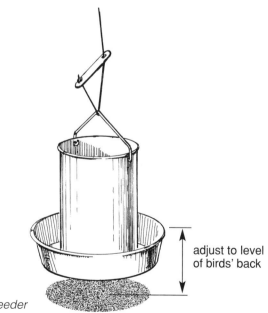

Range trough feeder with hinged roof

Hanging tube-type feeder

adjust to level of birds' back

Waterers

Birds of all ages should have access to plenty of clean, fresh water. Bear in mind that water is one of the cheapest sources of nutrients, and more than one half of the bird's body is water. Eggs are composed of about two-thirds water. Water is a vital ingredient for all of the bird's bodily functions.

Water consumption depends upon the environmental temperature, age, and species. Laying birds will drink about 2 pounds (0.9 kg) of water for every pound (0.5 kg) of feed consumed. In extremely hot weather, water consumption may amount to 4 pounds (1.8 kg) for each pound of feed eaten.

For young chicks the waterers can be commercial 1-gallon (3.8 L) glass or plastic jars with plastic bottoms. These are easy to clean. As the birds become older, exchange these for larger metal fountains or troughs. Water older birds in open pans, pails, or troughs. If the house has running water, a float-type valve can be installed in the waterline at the watering pan or trough so it fills automatically.

If running water is piped into the house, you may want to consider buying an automatic water fountain, cups, or nipple waterers. The cost is usually not excessive, the amount of labor required is reduced, and a constant supply of fresh water is available at all times. You may make a homemade waterer from a gallon fruit or juice can (see illustration). Waterers commonly used for small flocks of poultry are also shown.

Place a fountain on a platform or a water stand, covered with a 1 x 1 or 1 x 2 mesh welded wire. You can make the platform of 2 x 4 material 30 to 36 inches (76.2–91.4 cm) square. For best results, raise it 3 to 4 inches (7.6–10.2 cm) higher than the depth of the litter. Better yet, place it over a drain, if available. This arrangement will lessen the amount of wet litter surrounding the fountain and also help keep litter material out of the fountain. As with the feeders, the lip of the waterer should be level with the back of a standing bird. Adjust nipple waterers so the bird has to slightly stretch its neck to activate the waterer.

During cold weather in areas where freezing of water may occur, use heat lamps over waterers or electric pan heaters on which metal waterers can be placed. You can also use heat tape around pipes to prevent the freezing of waterlines. Remember, during cold weather, birds lose a lot of moisture just breathing and will require more water to keep in balance.

Waterers

Waterer made from a gallon fruit or juice can and a pan

Chick waterer   *Waterer for older birds*

Waterers (continued)

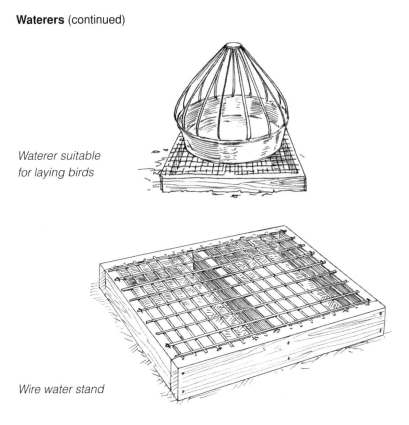

*Waterer suitable
for laying birds*

Wire water stand

Brooders

If you are starting day-old birds, you will need some type of brooder (heat source). You must provide enough brooder space so that every bird can get under the heat source as needed.

Brooders may be the hover type heated by gas, oil, electricity, wood, or coal, or infrared heat lamps. Infrared lamps work well for small flocks and have the advantage of enabling you to observe the chicks at all times. Electric brooders are more expensive to operate than most other types.

The first illustration is a plan for a homemade electric brooder for a small flock. The next shows an infrared brooder with guard. The typical hover type reminds one of a flying saucer. It is usually suspended from the ceiling by chain or cable.

Brooders

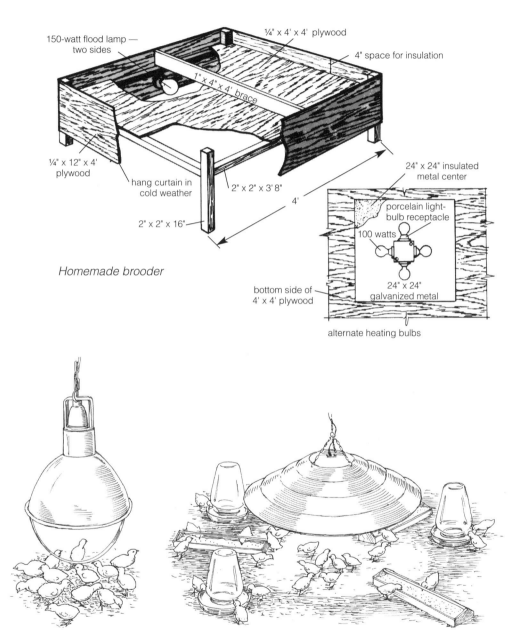

150-watt flood lamp — two sides

¼" x 4' x 4' plywood

4" space for insulation

1" x 4" x 4' brace

¼" x 12" x 4' plywood

hang curtain in cold weather

2" x 2" x 3' 8"

2" x 2" x 16"

4'

Homemade brooder

24" x 24" insulated metal center

porcelain light-bulb receptacle

100 watts

bottom side of 4' x 4' plywood

24" x 24" galvanized metal

alternate heating bulbs

Infrared brooder

Hover-type brooder

Roosts

Roosts are not essential but are recommended for most small laying flocks. Do not use them for broilers, roasters, or capons, as they may cause breast blisters. Use roosts for turkeys but not waterfowl. Allow pullets 6 to 8 inches (15.2–20.3 cm) of roost space per bird. If you expect the layers to roost, you must train them at an early age, for it will take considerably more time to train them to use the roosts in the laying house.

Dropping Pit

There are several types of roosts. One of the more common types is the dropping pit with perches on the top, which is designed to accumulate droppings underneath for several months. This, naturally, keeps a large amount of the moisture out of the litter. If you construct it properly and screen it in around the top and sides, birds will be prevented from getting into it. The nesting material and litter stay cleaner. Dropping pits are usually located either in the center or at the rear of the building — these are normally the more comfortable areas in the house. The dropping pit should be designed to facilitate catching and culling birds and so that it can be moved easily for cleaning. In larger housing units, the dropping pits should be narrow enough to facilitate easy catching or culling of birds at night.

Dropping pit with perches

Dropping Boards

Another roosting method uses a dropping board. The typical dropping board may be mounted on the wall at a height of about 2½ feet (0.76 m). The open space under the dropping board provides additional floor space for the birds. The roosts are located above the dropping board and the perches are hinged to the wall so they can be raised when the dropping board is cleaned. This method is not as desirable as the dropping-pit type, because of the difficulty of cleaning, and is not recommended unless floor space is at a premium.

The roosts or perches should be of 2 x 2 stock, rounded or beveled on the upper edges to prevent injury to the breast and feet; 1¼- to 1½-inch (3.2–3.8 cm) round stock is another option. For small breeds, allow 8 inches (20.3 cm) of perch space per bird. Large breeds should have 8 to 10 inches (20.3–25.4 cm) of perch space. Space roost perches 12 to 15 inches (30.5–38.1 cm) apart.

Nests

Provide an adequate number of well-designed nests with clean nesting material for the laying hens. They can be either individual or community nests. Provide one individual nest, or 1 square foot of community nest, for each four laying birds. Individual nests should be at least 1 foot (30.5 cm) square and 1 foot (30.5 cm) high. Community nests can be almost any size but must be provided with at least two 9-inch by 12-inch (22.9 by 30.5 cm) openings for every 20 square feet (1.8 sq m) of nest space. As an aid to getting the birds into the nests, darken each nest entrance by covering two thirds with a cloth flap. Place a landing board below the nest openings to provide easy access to the nests. Locate the nests approximately 2 feet (61 cm) from the floor or surface of the litter.

Commercially made nests are also available from agricultural supply houses and specialized equipment dealers. Some suppliers offer a community or individual type with a rollaway feature. After the eggs are laid, they roll out to the front or back of the nest into an egg tray, where they can be gathered easily. A word of caution about this type of nest: It's excellent in theory, but birds must be trained to use it and, more often than not, a large percentage of the eggs are laid on the floor rather than in the nest.

Nests

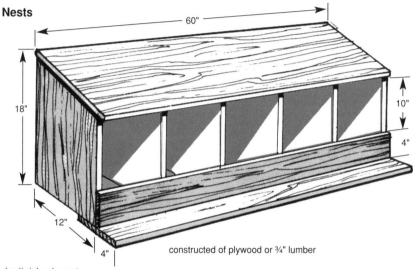

60"

18"

10"

4"

12"

4"

constructed of plywood or ¾" lumber

Individual nests

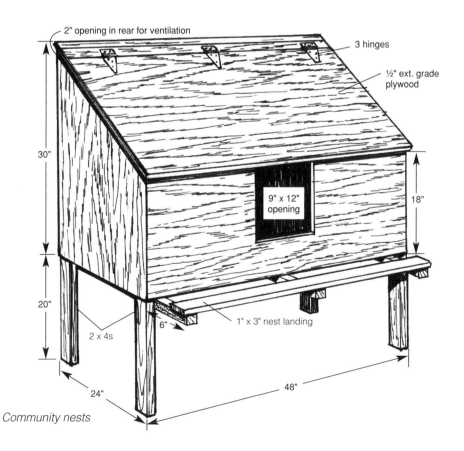

2" opening in rear for ventilation

3 hinges

½" ext. grade plywood

30"

9" x 12" opening

18"

20"

6"

1" x 3" nest landing

2 x 4s

24"

48"

Community nests

Clean the nests frequently, as the nest material becomes dirty. Because several hens use the community nest at one time, ventilation is important. For this reason, provide a 2-inch-wide (5.1 cm) opening at the top of the nest to allow heat to escape.

Cages

You may also house laying birds in wire cages. Locate the cages either in an open building or in closed, fan-ventilated housing. As mentioned earlier, cages simplify the management of layers, but birds in cages are more susceptible to the effects of extreme weather conditions. Therefore, you must protect them from wind, cold, and particularly from hot weather.

Cages are usually constructed of 1-inch by 2-inch (2.5 x 5.1 cm) welded wire; the floor is constructed with a slope of approximately 2 inches (5.1 cm) per foot, so that eggs will roll out onto the egg tray. Feed and water are provided in troughs usually located on the outside of the cage. You can also use water cups or nipples placed inside the cage.

There are many types of cages available, most of which are satisfactory. They come in several sizes but for best results should not exceed 20 inches (50.8 cm) in depth. Cages 12 inches (30.5 cm) wide, 18 inches (45.7 cm) deep, and 16 inches (40.6 cm) high are very popular. A minimum of 60 square inches (387 sq cm) per bird is recommended, no matter what type of cage is used.

The types of cages include the stair-step, the double-deck with dropping boards, the colony, and the single-deck. Stair-step cages are available in two arrangements, either full or modified. Most commercial producers use a triple-deck or four-deck modified stair-step cage system to enable them to house more birds in a given area.

The use of cages cannot be justified in many of the very small laying-flock enterprises; however, there are situations in which you can use them to good advantage. For example, cages can be used if you do not have a yard or sufficient space to properly range your flock. Consult your local extension agent or poultry specialist if you are not sure if cages would be of benefit to you.

Cages

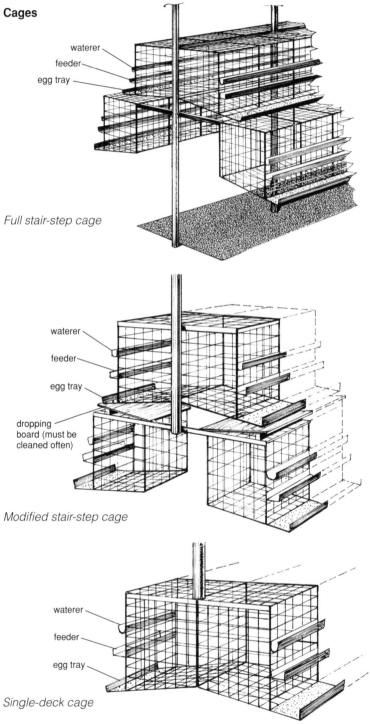

Full stair-step cage

Modified stair-step cage

Single-deck cage

BRANCHES, PONDS, AND BEACHES

Janette lives on a unique piece of property that has lots of trees, a small pond, and a stream running through it. She built several fenced-in flight cages with shade and roosting areas, and two buildings for feeding, laying, and roosting. She keeps more than one hundred birds, including ducks, geese, and chickens.

Janette started out small, with just a few Cochin and Old English bantams she purchased from a midwestern hatchery, but the flock has grown over the years. She became the collection point for stray and unwanted chickens, ducks, and geese from the area and, thus, her flock has increased in size a few weeks after Easter each year. Janette also finds new owners for some of her birds, placing them in what she refers to as "good homes." She makes sure the new owners have the facilities and the knowledge to properly raise chickens.

The ducks and geese have free run of her property during the day, and at night they go into a coop called South Beach. These waterfowl love the stream and the pond and sometimes all of them remain outside at night, preferring to sleep on the pond. Janette constantly worries about predators getting them because, despite her best efforts, over the years she has lost birds to hawks, coyotes, raccoons, and even free-roaming dogs.

Her chickens primarily have the run of two of the flight pens and the coop called North Beach. This is an old 20-foot by 40-foot (6.1 x 12.2 m) work shed converted to a maze of small pens that are interconnected with one another. The different breeds seem to find their own areas and claim them for roosting and feeding. Janette has placed tree branches strategically in the coop to serve as roosting sites and also provides nest boxes.

She maintains a litter floor of sawdust and shavings and keeps it clean and dry. She collects eggs every day, and her many friends and neighbors are recipients of an odd assortment of different-colored and different-sized eggs. Janette says that having the poultry around has kept her from being lonely since her husband died.

THREE

POULTRY
BIOLOGY

To better understand and manage your poultry flock, it is helpful to know a little about the biology of birds. This information will help explain why certain management techniques are called for.

This chapter provides some insights into the rearing and managing of poultry flocks. It is by no means meant to be comprehensive. The interested reader is encouraged to follow up with personal study of the many fine books on poultry biology, nutrition, physiology, and health found in most major libraries. Many good Web sites are available for the enthusiast and are listed in the appendix.

Biologically, chickens and other poultry are similar to mammals and other animals; however, there are some notable differences. For example, chickens have feathers instead of hair, wings instead of arms, no sweat glands, nonexpandable lungs with air sacs, and a body temperature between 105.6 and 107°F (41–42°C). Chickens have no teeth and use a muscular gland, the gizzard (ventriculus) to grind food. They have some characteristics in common with reptiles; for example, eggs, and scales on their shanks.

Our main interest in the chicken is as a supplier of food and, as such, we need to know as much about how it functions as possible, because in nature form and function are closely related. Most important, we need to know how to make it grow and reproduce efficiently.

Organ Systems

Poultry are complex animals with several organ systems, which, working together, make the animal function. These organ systems are:

- Nervous
- Skeletal
- Integument
- Digestive
- Urinary
- Reproductive
- Endocrine
- Exocrine
- Muscular
- Cardiovascular
- Respiratory
- Lymphatic
- Immune

It is most difficult to talk about one system without involving the others because they all work together to make the animal function. So the rest of this chapter focuses on some of the major organ systems and how they interact within a healthy bird. However, before we can fully understand the function of the organs and the whole animal, we must learn about its cells. Cells are what the organism is made of, and when we talk about feeding and caring for an animal, we are really talking about feeding and caring for all of its cells. If the individual cells that make up the organism are unhealthy or die, then it is difficult for the animal to survive.

Cell Biology

The cell is the basic structure of all simple and complex living organisms. Until the late 1830s, people knew nothing about cells. They thought that organisms just grew, sort of like crystals. Then Matthias Schleiden and Theodor Schwann proposed the cell theory, and from there it just blossomed. The invention of more powerful microscopes led to the discovery of the inner, working elements of the cell.

Two basic cell types have been identified. These are *eukaryotic cells* — those with a true nucleus — and *prokaryotic cells* — those without a defined nucleus (bacteria, blue-green algae).

Cells are composed of several subcellular systems, or organelles. The major organelles and other cell parts we know of are listed on page 40.

Major Organelles

- Nucleus
- Nucleolus
- Ribosomes
- Rough endoplasmic reticulum (RER)
- Smooth endoplasmic reticulum (SER)
- Golgi apparatus
- Lysosomes
- Peroxisomes
- Mitochondria
- Cytoskeleton
- Cytoplasm
- Vacuoles
- Cell membrane

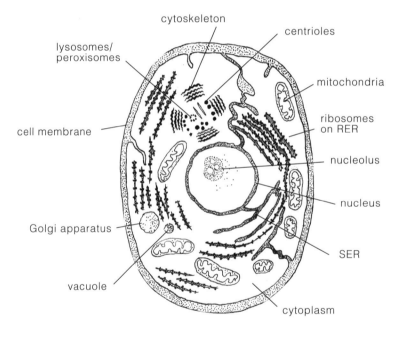

Anatomy of a cell

Nucleus

The nucleus holds the deoxyribonucleic acid (DNA) and is the largest organelle, occupying about 10 percent of the cell's volume. This is where the chromosomes and DNA are. The nucleus is the control center for many cell types. Some cells, such as muscle cells, have more than one nucleus.

Nucleolus

The nucleolus is in the middle of the nucleus and is the site of manufacture of the precursors of the ribosomes.

Ribosomes

The ribosomes read the genetic code from ribonucleic acid (RNA) and follow its instruction to make specific kinds of protein molecules. The protein molecules are the workhorse molecules of life that make up much of the structure of living organisms and, in the form of enzymes, carry out many of life's chemical processes. A cell that makes a lot of protein may have millions of ribosomes. Most ribosomes are outside the nucleus, stuck to the walls of a vast, irregularly shaped organelle called the RER. Some are also stuck on cytoskeletal fibers and synthesize proteins that are incorporated into cytoskeletal fibers, such as actin; or into the mitochondria, such as cytochrome c; and into the peroxisomes, such as catalase.

Endoplasmic Reticulum

This is a huge sac wrapped around the cell nucleus in an irregular pattern. There are many membranous channels between the folds of the sac. About half of the cell's volume is RER. The function of the endoplasmic reticulum is to provide a chamber, separated from the rest of the cell's metabolic processes, for the chemical modification of some of the proteins made by ribosomes. Ribosomes make some proteins for immediate use in the open liquid of the cell (cytoplasm). But if the protein is one whose function inside the cell cannot be fulfilled without being modified, the ribosomes adhere to the endoplasmic reticulum and inject the protein inside it. There, enzymes cause the chemical reactions that attract other molecules onto the proteins, enabling them to perform their function. Other proteins are partially modified and then encapsulated into a vesicle. The many ribosomes dotting its surface give the endoplasmic reticulum a rough appearance, hence its name.

Another type of endoplasmic reticulum, SER, is smooth and contains enzymes to process proteins and fats; it is the site of the synthesis and metabolism of fatty acids and phospholipids. The cells of the liver contain a lot of SER and RER.

Golgi Apparatus

Some proteins need further processing prior to leaving the cell. This happens in the Golgi apparatus, a medium-sized organelle. The number of Golgi in a cell depends upon the volume of secretory products of the cell. There are two regions of the Golgi, *cis* and *trans*. Proteins in little vesicles travel from the endoplasmic reticulum and attach to the cis region, pass through the Golgi, and exit through the trans region. Golgi are the "traffic police" of the cell, sorting many of the cell's proteins, tagging and directing them to the proper destinations.

Lysosomes

These small organelles are like the cell's stomach. They break down molecules of food that have been partially digested by the body's stomach. They contain about fifty digestive enzymes. These enzymes are so potent that if the membranes around the lysosomes were to rupture, the cell would be digested away by *acid hydrolysis*. Nutrients (from food) reach the lysosomes in vesicles that form at the cell's outermost membrane, the cell wall. The nutrient molecules are sucked into a pit that forms in the membrane. As the pit deepens into the cell, the membrane closes over it, forming a vesicle that breaks loose and travels deeper into the cell until it fuses with the lysosome's membrane.

During starvation of the animal (or cell), the lysosomes start to digest other organelles in the cell to feed the rest of the cell. Severe oxygen deprivation, such as in suffocation or drowning, causes an even more extreme reaction — cellular suicide. The lack of oxygen turns cells more acidic and the lysosomal membranes break down and release enzymes that destroy the cell from the inside. Brain cells are the first to do this, within about 5 minutes after breathing stops.

Peroxisomes

The peroxisomes are small organelles that contain enzymes that degrade fatty acids and amino acids and, in the process, produce hydrogen peroxide. So the peroxisomes also contain catalase, an enzyme that degrades the hydrogen peroxide to oxygen (O_2) and water (H_2O). A side product is heat, which helps maintain the proper temperature of the cell and the organism of which it is a part.

Mitochondria

The mitochondria are the third largest organelle in animal cells and occupy as much as 25 percent of the cell volume. They contain the enzymes that convert glucose plus adenosine diphosphate (ADP) to adenosine triphosphate (ATP) plus carbon dioxide (CO_2) plus H_2O, a process called *oxidative phosphorylation*. They also contain several copies of a small DNA molecule, mitochondrial DNA. This DNA codes for five or six of the key inner membrane proteins of the mitochondria. The ATP molecules produced by the mitochondria are used to power the cell. Cells can have as many as 5,000 mitochondria.

The Cytoskeleton

Lacing its way throughout the cell is a network of protein filaments of several kinds called the cytoskeleton, or the skeleton of the cell. Some are straight lines and some a filigree. Some maintain the cell's shape and some change to alter it. Some change continually to make the cell crawl or move. Others act as the framework of hairlike projections from many cells, called *cilia* or *flagella*, and cause them to wave or beat.

The cytoskeleton filaments most visible to the naked eye are made of the protein *keratin*. Hair and the outer layer of skin are made of keratin-rich cells that have died and lost most of their components.

One of the newest known functions of the cytoskeleton involves filaments called *microtubules*. Biologists have recently discovered proteins that act as motors, hauling vesicles over microtubule tracks. They have discovered a protein called *kinesin* that can pick up vesicles and transport them along microtubules in one direction. Another protein called *dynein* moves them in the opposite direction. How these "motor molecules" work is still unknown. It is speculated that the proteins hinge, attaching to the vesicle on one end and to the cell on the other, pulling each vesicle to the next set of "hinges," and so on, thus moving the vesicle along the cell until it reaches its destination. This discovery may help scientists uncover better ways to stimulate cell growth.

Cytoplasm

This jellylike fluid within the cells separates the organelles and allows substances to flow between organelles within the cell. It contains many of the proteins, molecules, and important substances that the cell needs, and also keeps the organelles separate. It is similar to air for humans: It has many essential ingredients for continued existence, and it provides a medium in which the organelles can exist.

Vacuoles

A vacuole is any membrane-bound organelle with little or no internal structure. It takes nothing from the cell and produces nothing for it. It does, however, store things for the cell. In a sense, it is a "vacuum." Vacuoles, though common in many cells, are most prominent in plant cells, taking up most of the central space.

Although the contents of the vacuole vary from organism to organism, as a rule they contain:

♦ Atmospheric gases
♦ Inorganic salts
♦ Organic acids
♦ Sugars
♦ Pigments

Cell Membrane

The cell wall and plasma membrane is thought to be a lipid bilayer surrounding the cell. It has both *lipophilic* (fat-attracting) and *hydrophilic* (water-attracting) properties, thus allowing it to transport fats and proteins across and into the cell. It also communicates with other cells, by certain recognition sites. It helps contact other cells to form *tissues*.

The exact control of movement of materials into and out of the cells is still being studied; however, many recent advances in cell biology have led to the discovery of beta-blockers to treat heart conditions and the drug lovastatin to reduce blood cholesterol levels. Further study of cell membranes may help scientists find ways to treat other conditions.

Extracellular Matrix

The extracellular matrix is what holds cells together to form organs. Much of this is called connective tissue, of which there are two kinds. The first is called *loose connective tissue* and is located between cells, where blood vessels and other structures are found, and its purpose is to hold cells together and feed them. The second type is known as *strong connective tissue* and is found in bone, cartilage, and tendons. Collagen is the major protein in the extracellular spaces of connective tissue. It is the most abundant protein found in the animal kingdom. It forms insoluble fibers that have a very high tensile strength and are secreted by *fibroblasts*.

There are many more structures and functions of cells, but this is not the forum for a lengthy discussion. If you are interested in learning more about cells and cell biology, visit your local library and find a recent book on the subject. This is a rapidly changing field of study, and new information is being gathered daily. Molecular biology is basically the study of the cells at the molecular level. Molecular and cell biologists have made the discoveries that have led to cloning and transgenics.

Animal Growth

One of the important questions in animal biology is: How does body growth occur in animals? The answer to this question is multifaceted.

Basically, there are two ways that growth occurs. One is called *hyperplasia*, which is an increase in the number of cells in the body, and the other is *hypertrophy*, which is an increase in the size of the cells. Each organ system of the body is made up of millions of cells, working to perform a common function. Muscles, for example, are designed for producing movement by contraction. Because muscles are also the economically important tissue of broiler chickens and other meat birds, the following discussion first concentrates on how muscles grow and function.

Muscle Growth

There are three basic types of muscles in animals: skeletal, smooth, and cardiac. Smooth muscles are found in blood vessels and intestines. The function of muscles is primarily for the movement of the body and, in the process of movement, to generate heat. They also fill out the contour of the body and give each animal part of its characteristic shape. After puberty, most muscle growth is due to hypertrophy.

The pectoral muscles, or breast muscles (flight muscles), are the largest in the bird and are therefore the most economically important, providing more than half the edible meat of the chicken. The gastrocnemius muscle on the back of the leg is attached to the toes by the gastrocnemius tendon, which when relaxed causes the toes to grasp a limb, allowing a bird to sleep on a perch.

Two major types of skeletal muscle are of importance: light and dark muscle (also known as white and red muscles). Light muscle is such because of its low level of *myoglobin,* the oxygen-holding molecules in muscles, and because it has a definite fibrillar appearance, whereas red fibers have a more granular appearance and more myoglobin. Muscles that are used more, such as leg muscles and flight muscles in wild birds, tend to accumulate more myoglobin. Because most domestic chickens and turkeys are genetically selected and raised for meat production and don't fly much, their breast muscles are light. Broiler chickens have muscle fibers that are larger in diameter and lighter in color than those of layers, mainly due to genetic selection.

Proper muscle development and growth depends upon the proper nutrients being supplied to the chicken. In the embryo, this means yolk, white, and eggshell nutrients, plus the proper temperature and humidity. In hatchlings and young birds, this means a balanced ration and clean water. The proper balance of amino acids and minerals is the key to proper muscle growth. Muscles also need calcium, in addition to protein, to function properly. The point is, keep the birds fed properly and muscle growth will be normal. The energy source to keep muscles working comes mainly in the form of a carbohydrate called *glycogen.* Glycogen is stored in muscle tissue and the liver and is utilized as needed, such as during exercise or when blood glucose levels fall. Storage of glycogen is dependent upon the pancreas and its release of insulin (see the section on endocrine glands, page 60, for more on this issue).

Other Cell Growth

Other cells of the body, such as blood cells, feather-follicle cells, and skin cells, always grow by cell multiplication; that is, a progressive increase, such as 1, 2, 4, 8, 16, 32, and so on. Growth potential is never fully reached, because of regulation of growth by hormonal or genetic control.

Nerve tissues and muscle (after hatching) grow by cell enlargement (hypertrophy). This same process occurs in fat cells once puberty is reached. Following feeding, lipids are mostly stored in existing adipocyte vacuoles, rather than in newly formed cells.

Fat

Body fat, to a large extent, acts as an energy reserve and is the most variable among the major body constituents. While it varies with species, sex, and age, it is also strongly affected, quantitatively as well as qualitatively, by nutrition. Even following prolonged starvation, body fat is never completely depleted and is not likely to drop below 4 percent of total body weight, because of the need to protect the integrity of tissues and organs.

A laying hen fed a commercial laying diet will absorb little more than 3 grams of fat per day. Because an average egg yolk contains approximately 6 grams of fat, an appreciable part of the yolk lipid of a hen laying an egg almost every day must be synthesized from nonlipid constituents. Much of this comes from carbohydrates and, to a lesser extent, from proteins. Thus, absorption of fats from the intestinal tract and the synthesis of fats from nonlipid compounds become two distinct sources of depot (body fat), organ, and egg lipids. About 2.5 percent of the fat in chickens is found in the abdomen. By the way, whole chickens purchased at the supermarket in which this fat has been removed can be labeled low-fat chickens according to U.S. Department of Agriculture (USDA) rulings.

Recently, chemicals called partitioning agents have been tested that direct nutrients into the development of muscle and skin tissue. This may increase the protein content of the body by as much as 10 percent while decreasing fat by as much as 25 percent. None has been approved for general-use poultry at this time.

Bones

Bone growth is most rapid after hatching, followed by muscle growth, with fat accumulation the least rapid in growth. Bone growth is dependent upon the proper levels of calcium, phosphorus, other minerals, vitamin D, parathyroid hormone, calcitonin, growth hormone, and steroids (estrogen in females and androgen in males). For example, the shanks of female chicks reach full growth in 16 to 18 weeks, whereas mature body size is not reached until 40 to 50 weeks after hatching. But the long bones of males, controlled by all of the above factors, including genetics, will be larger than those of the female. Remember that without proper skeletal development, muscle development will not occur properly, either. Tibial dyschondroplasia, or leg weakness, is a problem with some rapid-growing birds. It occurs as a result of an overabundance of chondrocytes, cartilage cells, and not enough osteocytes (bone cells), due to a possible surge in growth hormone. The new bone growth occurs without proper mineralization and vascularization, so the bone is soft and unable to support the animals' weight.

Each bone has its special period of growth. The last bone to reach full maturity is the posterior part of the sternum (keel). A bone grows in length because of *cell division* (increase in the number of cells). The dividing cells are called *osteocytes* and are found in the *epiphysis,* an area just below the end of the bone. On the other side of the epiphysis, toward the joint, new osteocytes are made. As they mature, they are surrounded by a matrix of connective tissue that helps keep the cells in the proper arrangement. At the side of the epiphysis away from the joint, the osteocytes change and begin to calcify. As this process occurs, the matrix is reabsorbed. During this process, both the diameter and the length increase until puberty.

Females have specialized bone called *medullary bone,* which is used as a storage area for calcium for egg formation. The medullary bone is influenced by parathyroid hormone, calcitonin, vitamin D, and estrogen. This bone is formed about 10 days prior to the

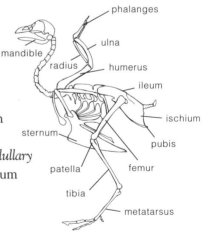

Skeletal anatomy of a chicken

formation of the first egg and results in about a 10 percent increase in skeletal weight. This occurs only in laying hens, not in males or immature females. Medullary bone is found in the tibia, femur, pubic bone, sternum, ribs, ulna, toes, and scapula. This type of bone fills the marrow cavity with fine interlacing spicules of bone to provide a ready supply of calcium.

The skeleton of birds is compact, lightweight, and very strong. The vertebra of the neck are flexible, but the thoracic and lumbar vertebrae are fused to make a solid, strong support structure for flying birds. The sternum has the flight muscles attached to it and protects the vital organs.

Chickens also have specialized bones called *pneumatic bones* that are connected to the respiratory system through the air sacs. These bones are hollow. They include the skull, humerus, clavicle, keel, and the lumbar and sacral vertebrae. A chicken could actually breathe through a broken humerus.

Integument

The comb, wattles, feathers, and preen gland and skin are all part of an organ called the integument. The comb, wattles, and preen gland are considered specialized skin. The scales on the shanks are also special epithelial (skin) cells. The chicken has no sweat glands in the skin. Thermoregulation occurs by other means to be discussed later. The comb and wattles are used in identification among birds and as an area for dissipation of excess body heat.

Feathers

When the chick hatches, it has almost no feathers. Except for the wings and tail, it is covered with down. The down soon grows longer, develops a shaft, and becomes the first set of feathers for the chick. By the time the bird is 4 to 5 weeks of age, it is fully feathered. These first feathers soon molt and a new set is grown in by the time the bird is 8 weeks old. The third set is completed just prior to sexual maturity.

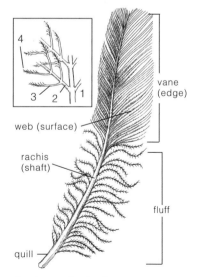

Anatomy of a feather
(*1=shaft; 2=barb; 3=barbule; 4=barbicel*)

Feathers make up from 4 to 8 percent of the live weight of the bird. Older birds and males have the lowest percentage. Feathers cover the body in specific tracts. There are ten feather tracts: head, neck, shoulder, breast, back, wing, rump, abdomen, thigh, and leg. Feathers are used to protect the bird, aid in flight, and insulate the bird, as well as help distinguish between the sexes.

The order of feathering is as follows: shoulder and thigh, 2 to 3 weeks; rump and breast, 3 to 4 weeks; neck, abdomen, and leg, 4 to 5 weeks; back, 5 to 6 weeks; and wing coverts and head, 6 to 7 weeks.

The feather is made up of the quill, which is carried through the vane as the shaft or *rachis*. The barbs branch from the quill and the barbules branch from the barbs. The barbicels, which branch from the barbules, have small hooks on the end, called the hamuli, that are used to interconnect with other barbicels to make the feather more airtight. This attachment is similar to how hook-and-loop tape works.

Slow feathering is a dominant trait and is sex-linked. However, fast feathering is desirable on broilers and laying hens. In the rapid-feathering chick, the primary and secondary feathers are longer than the coverts. In slow-feathering chicks, the primaries and coverts are about the same length at hatching. The cross between a fast-feathering male and a slow-feathering female results in a slow-feathering male and a fast-feathering female.

Molting, the regression of the reproductive tract and loss of feathers, is important in layers and breeders, as it extends the laying life of the bird.

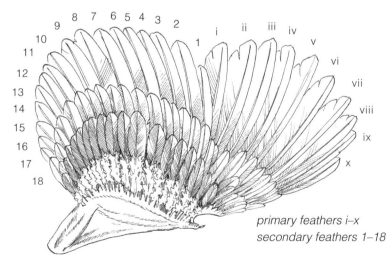

primary feathers i–x
secondary feathers 1–18

Molting order of secondary feathers: 11, 12, 13, 14, 15, 16, 17, 18, 10, 2, 3, 4, 5, 6, 7, 8, 9, 1.

Feather loss during a molt is in a definite pattern starting with the head and proceeding to the neck, body, wings, and tail. In a normal, natural molt, the primary flights are lost at about one per week, with the first to be shed being the one next to the axial feather. It takes about 6 weeks to grow out a new feather. The secondaries molt in this order: 11, 12, 13, 14, 15, 16, 17, 18, 10, 2, 3, 4, 5, 6, 7, 8, 9, 1. (See chapter 6 for more on molting.)

Skin

The skin of the bird is thin over much of the body, except for the comb, wattles, earlobes, and preen gland. The preen gland stores fats, and upon irradiation by ultraviolet light it forms vitamin D_3, which is worked into the skin by rubbing it with the beak. Yellow skin is the result of the hydroxycarotenoid pigment *xanthophyll,* from the feed. Corn, alfalfa, and marigold petals are sources of xanthophyll. Yellow yolks are also a result of the pigment, which is lost in parts of the skin as hens lay eggs. The number of eggs necessary to bleach various parts of the body are given in the chart. As the bird lays eggs, all the pigment is diverted into the yolk so it leaves the body; but when the hen stops laying eggs, the pigment returns in the reverse order.

Xanthophyll and Skin Bleaching*

BODY PART	NO. OF EGGS REQUIRED TO BLEACH
Vent	1 egg
Eye ring and earlobes	9–12 eggs
Beak, starting from near the face to the tip as follows:	
Inner ⅓	11 eggs
Inner ½	18 eggs
Inner ⅔	23 eggs
Inner ⅘	29 eggs
Whole beak	35 eggs
Bottom of feet	66 eggs
Front of shanks	95 eggs
Back of shanks	159 eggs
Tops of toes	175 eggs
Hock joints	185 eggs

*Note: *Xanthophyll is a yellow pigment that is lost from the skin as hens lay eggs*

Cardiovascular System

The function of the blood is primarily to transport oxygen to the tissues and to remove carbon dioxide from them; it is secondarily to transport other nutrients to the tissues and to remove waste products from them to be disposed of by the kidneys. Blood also transports the hormones from the endocrine glands to their target organs. With the aid of the kidney, the cardiovascular, or circulatory, system helps regulate body water. The antibodies are also transported in the blood. Blood constitutes about 5 percent of the body weight of the chick and about 9 percent of that of the adult chicken. It contains a fluid portion (*plasma*), salts and other chemical constituents, and certain formed elements, the *corpuscles*. The corpuscles include the *erythrocytes* (red blood cells) and *leukocytes* (white blood cells). The erythrocytes of birds are oval in shape and, unlike those of mammals, are nucleated and larger. In adult chickens, there are about 3.0 (female) to 3.8 (male) million red blood cells per milliliter of blood. The leukocytes, which function in immunity, are of several types. There are heterophils, eosinophils, basophils, lymphocytes, and monocytes. Leukocyte counts are used in determining the level of immunity a bird has.

Heart rate varies from about 280 to 340 beats per minute in adult chickens, with a blood pressure of about 166/133 (systolic/diastolic). The heart is a muscle made up of specialized tissue. The muscle fibers, called *cardiac muscle,* are unique to the heart. The cardiovascular system works in conjunction with the respiratory system. The lymphatic system is a ductless system that eventually dumps lymph into the blood at the thoracic duct.

Respiratory System

The respiratory system comprises the nasal cavities, pharynx, superior larynx, trachea, inferior larynx, bronchi, lungs, and air sacs, and also the hollow bones. The air sacs are divided into one pair of abdominal air sacs, one pair of cervical, two pairs of thoracic, and one single interclavicular air sac. Chickens do not have diaphragms. Breathing occurs when the abdominal muscles relax, the inspiratory muscles contract, and the body cavity expands, thereby creating a negative pressure in the air sacs,

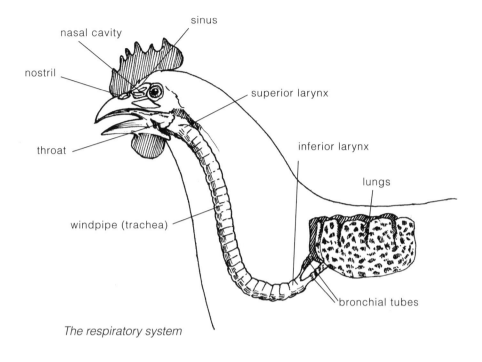

nasal cavity

sinus

nostril

superior larynx

throat

inferior larynx

lungs

windpipe (trachea)

bronchial tubes

The respiratory system

drawing air in through the lungs and over the parabronchi. Upon expiration, the expiratory muscles contract and force air back through the lungs and out the mouth and nostrils. The air sacs function similarly to bellows, with their pressure and volume modulated by the muscles of respiration. The avian lung is a rigid system of air and blood capillaries in intimate contact. The lung does not expand like the lungs of mammals and is actually a "flow-through system" with a small gas volume that does not expand to accept the tidal volume.

The air sacs and lungs also permit the evaporation of moisture and, with it, latent heat. This is one of the major ways in which chickens thermoregulate. Additionally, the air sacs perform the following for the bird: act as a bellows moving air across the gas-exchange surface of the lungs, as a balloon to lower the specific gravity, as ballast to obtain the proper center of gravity, as a friction pad between muscle action, and as a reservoir of air during violent muscle contractions.

The pneumatic bones are connected to the respiratory system through the air sacs. The air pressure makes the pneumatic bones light and rigid.

Digestive System

Chickens are omnivores; thus, their digestive system has characteristics of both herbivores and carnivores. They have no lips, soft palate, cheeks, or teeth, but do have an upper and lower horny mandible to enclose the mouth. The mandibles are referred to as the *beak*. The upper is attached to the skull, while the lower is hinged. The *hard palate* (roof of the mouth) is divided by a long, narrow slit in the center that allows air flow to the nasal passages. Because of this, the bird is unable to create a vacuum to draw water into its mouth. It drinks with the aid of gravity, which is why it raises its head every time it drinks water. The tongue is pointed and has a rough, brushlike surface at the back to help push food particles down the esophagus. Saliva, with the enzyme amylase, is secreted in the glands of the mouth to help lubricate food passing down the esophagus to the crop.

The *crop* is an enlargement of the esophagus, found just prior to where the esophagus enters the thoracic cavity, which acts as a storage place for food before it is sent down to the *proventriculus*, or true stomach.

In the proventriculus, hydrochloric acid and enzymes, such as pepsin, are added to the food to aid in the digestive process. After the

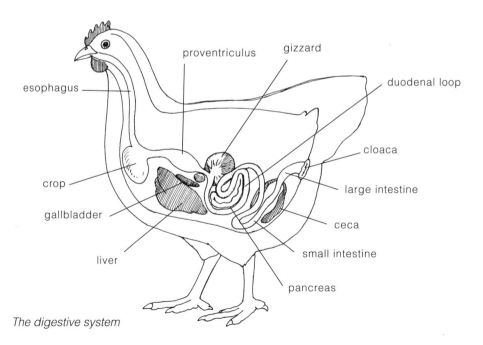

The digestive system

proventriculus, the food passes into the ventriculus, or gizzard, where the food is mixed with the acid and enzymes from the proventriculus and ground into smaller particles. The addition of grit to the food will aid in this process, though it is not absolutely necessary. The time spent in the gizzard depends upon the coarseness of the food — the finer the particles, the less time spent.

After the gizzard is the small intestine, the first part of which is the duodenum. Here, the food is mixed with secretions from the liver and the pancreas. The bile, stored in the gallbladder, is manufactured in the liver and contains enzymes and salts to aid in the digestion of fats. Cholesterol is also contained in the bile and is reabsorbed by the small intestines. Pancreatic enzymes actually help neutralize the acid from the proventriculus and aid in the digestion of starches, fats, and proteins. The pancreas secretes five different enzymes into the duodenum, including amylase, lipase, and trypsin. Other enzymes are produced in the wall of the small intestine and further the digestion of protein and sugars.

Absorption of the nutrients from the intestine to the bloodstream takes place in the duodenum and small intestine and is completed prior to their entering the large intestine.

At the point where the small and large intestines meet, the *ceca* are found. These are two blind sacs in which some bacterial digestion takes place, but relatively little of their function is yet known.

The large intestine is only about 4 inches (10 cm) long and about twice the diameter of the small intestine. It is a place for water reabsorption to take place and helps maintain water balance in the bird. The terminal portion of the large intestine is called the *cloaca*. It is a bulbous area at the end of the alimentary tract and means "common sewer." The digestive, urinary, and reproductive tracts all empty here. The *vent* is the external opening of the cloaca.

The liver, in addition to functioning as part of the digestive system, is also the site of blood detoxification and storage of glycogen and vitamins for use by the tissues as needed. The liver also manufactures uric acid, which is filtered by the kidneys. Uric acid is the end product of protein metabolism, and the form in which excess nitrogen is excreted. The acid is mixed with the feces in the cloaca and is the white product seen in bird droppings.

Urinary System

The uric acid produced by the liver is filtered by the kidneys and excreted into the cloaca via the ureters. Chickens have no urinary bladder and do not store urine. The kidney basically has three functions: filtration, excretion or secretion, and absorption. It filters water and some substances normally used by the body from the blood, along with the waste products of metabolism that are voided in the urine. It conserves needed body water, glucose, sodium, and other substances by reabsorption. The kidney also produces the enzyme renin, which helps regulate blood pressure, and synthesizes metabolites of vitamin D to aid in the absorption of calcium for shell and bone formation.

Reproductive System

One of the main objectives of each living species is to perpetuate itself through reproduction. Reproduction is accomplished in many ways, but all methods fall basically into two categories: asexual and sexual. Where individuals reproduce by asexual means, they do so without the union of male and female germ cells. They are able to subdivide to produce two individuals and subsequently more. Sexual reproduction requires the union of male and female germ cells to produce a new individual.

Birds reproduce by sexual means and, unlike most other animals, reproduce by means of an egg. The egg contains all the necessary elements to support life processes during embryonic development. The egg components and the developing embryo are protected by a strong shell.

The reproductive organs include the ovary and oviduct in the female and the testes in the male. The reproductive system of birds is somewhat unique in the animal kingdom. Most avian females develop only one functional ovary and accompanying oviduct. Still, with just one ovary, females of many avian species can be induced to lay eggs on an almost continuous basis. Truly, this is a remarkable system. For the modern poultry industry, production *is* reproduction; therefore, knowledge of the reproductive process is of tremendous importance to the successful producer. Understanding how the reproductive process works can help producers better manage their birds for optimum egg production.

Males

The testes of the male bird are paired but, unlike those of most mammals, are located within the body cavity, near the tip of the kidneys. Thus, sperm is produced at body temperature in birds, not a degree or two lower, as in mammals. The weight of the testes in chickens constitutes about 1 percent of the total body weight at sexual maturity. This is quite large. They are almost kidney-bean-shaped, and a grayish yellow color. In immature chickens, they are much smaller and yellow.

Internally, the testes are made up of a dense series of convoluted seminiferous tubules, connective tissue, blood capillaries, and Leydig cells. In the seminiferous tubules, the sperm are formed from germinal tissue. The Leydig cells produce the male steroid hormone testosterone. Testosterone is responsible for maturation of the sperm cells and for secondary sexual characteristics of males, such as feather patterns, colors, and crowing in chickens.

The accessory sexual reproductive organs of the male include the vasa efferentia, epididymis, ductus deferens (the tube system from the testes to the urodaeum of the cloaca), and the ejaculatory groove and phallus. In the process of mating, intromission does not occur in the chicken or turkey. The semen is transferred to the cloacal area near the phallus, the engorged phallus contacts the everted cloaca of the female, and sperm is then transferred to the female, in what is called the *cloacal kiss*. The phallus of the drake and other species of birds, however, is a vascularized, spiral-shaped sac with an ejaculatory groove that can protrude as much as an inch (2.5 cm) or so.

The average male chicken produces about 1 mL of ejaculate containing as many as 3.5 billion sperm. This large volume is produced because sperm can live within the hen for up to 20 days after mating and are needed to maintain fertile egg production for that time.

Light is the major factor controlling sexual maturity and sperm production in the male. Follicle-stimulating and luteinizing hormone (FSH and LH, respectively) are released from the pituitary in response to increasing day length and, in turn, stimulate the testes to produce sperm and testosterone.

Females

The right ovary and oviduct are present in embryonic development in all birds, but only the left ovary and oviduct develop in Galliformes (fowl). Among falcons and brown kiwi, however, both the left and the right ovaries and oviducts are commonly functional. In sparrows and pigeons, about 5 percent may have two ovaries.

In the chicken, the left ovary of the immature bird consists of a mass of small ova, of which almost 2,000 are visible. In a mature, functioning ovary, a hierarchy of developing follicles (yolks) is found, with between four and six large yolk-filled follicles seen at any time. The largest one is the next to be ovulated to produce an egg. The follicle is covered by a membrane that is highly vascularized, except for one section, called the *stigma*. The follicular membrane ruptures along the stigma at ovulation. If the follicular membrane ruptures anywhere else, a small amount of blood will be released along with the follicle (yolk) and produce a blood spot in the egg.

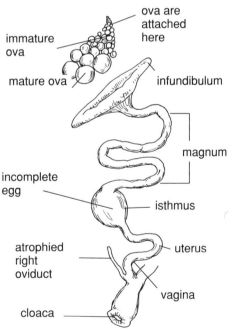

After ovulation, the egg is caught in the first section of the oviduct, called the *infundibulum*. In the infundibulum are sperm-storage areas, and this is where fertilization takes place. The first layer of albumen, called the *chalaza*, is also produced here. It takes about 15 minutes for the egg to pass through the infundibulum.

The yolk next passes to the *magnum*, the longest part of the oviduct, where most of the albumen is formed and deposited around the yolk. It takes about 2 to 3 hours for the egg to pass through the magnum.

Following the magnum is the *isthmus*. Here, the inner and outer shell membranes are formed around the egg, and it takes about 1¼ to 1½ hours to pass through.

The female reproductive system

The egg then enters the shell gland, or *uterus*. Here, the egg takes up salts and water in a process called *plumping,* and the hard shell is added. The egg spends about 18 to 21 hours in the uterus, and this is where final calcification of the shell takes place. Following the uterus, the egg enters the vagina. The cuticle layer is added over the shell as it passes into the vagina, where the egg turns 180 degrees, so that the large end comes out first.

The ovaries are responsible for production of estrogen and progesterone in the female. Estrogen stimulates secondary sexual characteristics and aids in calcium metabolism in the hen.

Nervous System

The nervous system is divided into the *central nervous system* (CNS), which includes the brain and the spinal cord, and the *peripheral nervous system* (PNS), which includes the cranial and spinal nerves, which emanate from the brain and the spinal cord, respectively, and the many ganglia and plexi that are associated with visceral innervation.

The nervous system is also divided by the type of control it provides, with the *somatic* being under normal conscious control, such as the voluntary muscles of the skeleton, and the *autonomic* being under reflexive or normally unconscious control, such as the viscera. Nerve cells, or *neurons*, are in contact with other neurons, through a synapse, to form a vast neural network constituting the nervous system. Neurons have a cell body and an axon that sends impulses away to another neuron, and dendrites that receive impulses from another neuron. Specialized neurons, called *sensory neurons*, act to sense the environment for animals. They help perceive heat, cold, light, sound, position, taste, smell, pressure, and pain.

Vision

The avian eye is similar to vertebrate eyes but with many adaptations that have improved its visual abilities. It is considered the finest ocular organ among animals. Most birds can focus within a range of 20 diopters, twice that of a 20-year-old human. Their color vision peaks in the orange-red portion of the spectrum, and they have both monocular and binocular vision with a 300-degree field of view.

Hearing

Birds have a keen sense of hearing and a high degree of equilibration. Their excellent voice production and remarkable ability to imitate sounds have implied an exceptional degree of sound analysis (pitch discrimination) within a wide range of auditory frequencies.

Taste

Chickens are indifferent to common sugars, but have a specific taste for sodium chloride. They will not eat food with salt concentrations exceeding 0.9 percent. This is about the limit their kidneys can safely handle. They have a range of acceptance for sour, either acid or alkaline, and show a sensitivity for bitter taste. Another interesting fact is that chickens will not drink water that is about 10 degrees hotter than their body temperature, but will drink water down to freezing.

Endocrine Glands

The endocrine glands are ductless glands that secrete chemical substances (hormones) into the blood. They act on specific target organs. The glands are:

- Pituitary-hypothalamus complex
- Gonads
- Pancreas
- Adrenal glands
- Thyroid glands
- Parathyroid glands
- Ultimobranchial glands
- Pineal gland

In addition to these "classic" endocrine glands, hormonelike substances are also produced by:

- Endocrine cells of the gut
- Thymus
- Liver
- Kidney

We will discuss each gland and its secretions and their effects on the target organs and on the entire body.

Pituitary Gland

The pituitary has been called by many the master gland of the body. It is also known as the *hypophysis* and in combination with the hypothalamus is known as the hypothalamic-hypophyseal complex. The hypothalamus is largely responsible for releasing factors to stimulate the pituitary to release its hormones.

The pituitary gland is divided into the anterior and the posterior pituitary. The posterior pituitary gland consists of the neurosecretory terminals, which release *mesotocin,* also called arginine vasotocin. Mesotocin is released to cause the bearing-down reflex, or oviposition (laying), of the egg. It is also the major antidiuretic hormone in birds.

The anterior pituitary secretes several types of hormones: gonadotropins, thyrotropin, prolactin, somatotropin, adrenocorticotropic hormone, and related peptides.

Gonadotropins

The gonadotropins we are most interested in are LH and FSH. These are both glycoprotein hormones. In the male chickens, as in mammals, LH acts primarily to stimulate the Leydig cells of the testes to differentiate and to produce testosterone, while FSH promotes spermatogenesis.

Both LH and FSH are necessary for normal ovarian physiology; however, their exact roles have not been fully established. Ovulation is induced by LH, and both LH and FSH stimulate ovarian production of several steroid hormones. LH has the major effect on the largest follicles to produce estrogens, testosterone and progesterone, while FSH works on the granulosa cells of the smaller follicles. The secretion of LH and FSH is controlled by hypothalamic releasing factors.

As levels of estrogen increase in the female, it causes a negative feedback on the pituitary to stop the release of LH and FSH. Progesterone, on the other hand, exerts a positive feedback on the release of LH.

There is a preovulatory surge of LH from 4 to 6 hours before ovulation; FSH has been found to increase about 14 hours prior to ovulation.

Thyrotropin

Thyrotropin, or thyroid-stimulating hormone (TSH), affects many aspects of thyroid function in the chicken, stimulating growth and the production and release of thyroid hormones. Thyroid-stimulating hormone has a number of effects on thyroid cells that eventually lead to thyroxine (T_4) and triiodothyronine (T_3) secretion into the bloodstream. There is also evidence that TSH is released in response to the secretion of thyrotropin-releasing hormone (TRH) from the hypothalamus.

Prolactin

In birds, prolactin is related to either reproduction or osmoregulation and possibly to the regulation of metabolism and growth in some birds. Prolactin stimulates the production of crop milk in pigeons and doves. In other avian species, prolactin is related to incubation behavior (broodiness) and to inhibition of ovarian functioning. The exact relationship between prolactin and broodiness has yet to be elucidated.

Somatotropin

Somatotropin, or growth hormone (GH), has been isolated from the pituitary of many species of birds, including chickens. The lack of GH due to removal of the pituitary gland (hypophysectomy) results in a dramatic decrease in the growth rate of chickens. Growth hormone exerts short-term effects on metabolism in birds and has a lipolytic effect, meaning it will cause the utilization of fats and lipids as an energy source. This is important in the broiler industry, with the search for fast-growing, lean birds. Growth hormone also decreases lipogenesis, the increased manufacture and deposition of fat, and reduces glucose utilization. During periods of nutritional deprivation, GH levels are increased and result in the above-mentioned actions.

The release of GH is controlled by several factors: (1) TRH, (2) a GH-releasing factor (GRF), and (3) somatostatin, which inhibits GH release. Growth-hormone release is inhibited by prostaglandins, and by insulin and glucagon in young chickens. It is also inhibited by nutritional deprivation, such as 1 or 2 days of starvation, and by chronic feed restriction, as might occur during molt.

Aberrantly low levels of thyroid hormones will cause an increase in GH release. Growth hormone in chickens is still under study.

Adrenocorticotropic Hormone

Adrenocorticotropic hormone (ACTH) acts to stimulate corticosterone production by the cortical cells of the adrenal gland and is controlled by corticotropin-releasing factor (CRF).

Gonads

The gonads in fowl consist of the testes in the male and the ovary and oviduct in the female. Male and female fowl respond to light in a similar manner. Increasing the daylight length causes a release of gonadotrophic hormones from the anterior pituitary. In the case of the male, there is an increase in the size of the testes, androgen secretion, and semen production, along with a stimulation of the mating behavior. In females, the ovarian follicles are stimulated by FSH from the anterior pituitary gland. This causes the ovary to develop and sets into motion a series of events, culminating in the development of the oviduct and preparing it to be able to form and lay eggs.

The hormonal secretions of the testes and ovary are important for the development of secondary sexual characteristics of birds and mammals as well as for the synchronization of sperm and egg development and mating between the sexes. The hormones also control nest building and brooding behavior in the female and mating and protective behavior in the male. Without hormones, reproduction and production would come to a halt.

Testes

The testes produce testosterone, which has a negative feedback effect on gonadotropin secretion. Additionally, relatively large quantities of progesterone and smaller quantities of estrogen (estradiol 17-B) have been found in testicular tissues. The Leydig cells appear to be the major source of this hormonal activity.

Androgens affect secondary sexual differentiation such as comb size, plumage, bill color, feather structure, vocalizations, and behavior.

Ovarian Hormones

Plasma concentrations of progesterone and estrogen have been shown to increase about 6 to 8 hours prior to ovulation in the largest and smallest preovulatory follicles, respectively. Production of progesterone by the ovarian granulosa cells is greatly enhanced by LH release. Both FSH and LH stimulate estrogen production by the thecal cells.

Progesterone may inhibit ovulation or induce molt when administered in large doses.

Ovulation is stimulated by an as yet unknown mechanism, though ovulation usually follows egg laying within 15 to 75 minutes. Blood LH peaks about 4 to 6 hours prior to ovulation; FSH peaks about 15 hours prior to ovulation.

Estrogens have also been implicated in the regulation of calcium metabolism in the hen, along with vitamin D_3 and calcium-binding protein. It is believed that estrogen stimulates the production of calcium-binding proteins.

The hen's reproductive cycle is influenced by photoperiod, with decreasing day length being destimulatory and increasing day length being stimulatory. (*Photoperiod* is the length of the light period to which birds are exposed to permit the best results in terms of growth, egg production, or semen production.) It has been found that less than 12 hours of light will result in less than maximal production, whereas 12 to 18 hours of light per day will result in normal egg production. It has been shown that the amount of darkness, and when it is applied, will influence oviposition time. This has been demonstrated using intermittent lighting.

Ahemeral photoperiods (shorter or greater than 24 hours) have also been shown to affect egg production. A photoperiod of 27 hours reduces egg numbers but increases egg size.

Pancreas

The pancreas is located in the abdominal cavity, in the loop of the duodenum. Much of the organ is devoted to the synthesis of digestive enzymes: proteolytic, lipases, and amylolytic enzymes. About 1 to 2 percent of the pancreas is dedicated to the synthesis of the endocrine products: insulin (anabolic), glucagon (catabolic), pancreatic polypeptide, and somatostatin.

Insulin and glucagon are primarily involved in the regulation of blood-sugar levels. Briefly, insulin stimulates the production of glycogen from glucose and its storage in muscle and liver tissue. Glucagon works to increase *gluconeogenesis*, or production of glucose, from glycogen and other sources, along with corticosterone.

Somatostatin appears to help in the regulation of the absorption and distribution of recently digested nutrients, as well as in the paracrine regulation of glucagon and insulin.

Adrenal Glands

The avian adrenal glands are located at the tip of the kidney. The glands are divided into the outer cortical tissue and the inner medulla. Corticosterone, a glucocorticoid, is the principal steroid hormone of the adrenal cortex. Aldosterone, a mineralocorticoid, is the other major hormone released by the cortex. Corticosterone has been called the stress hormone, because it is released in large concentrations when the bird receives stimuli it perceives as a threat to its survival or well-being.

Corticosterone has many physiological effects in animals. Although it may cause an increase in food intake, it likewise reduces growth in birds. The reduced body weight is accompanied by an increased deposition of body fat. This is caused by increased lipogenesis by the enlarged liver and by changes in lipolysis. Corticosterone acts to increase net muscle-protein breakdown and plasma glucose and liver glycogen concentrations by stimulating the production of glucose from these stores of carbohydrates in the body.

Birds that are under stress are more prone to sickness and disease. This is because corticosterone released in response to stressors is an immunosuppressant. The longer the period of stress, the greater the chance the bird will succumb.

Aldosterone secretion is controlled by the renin-angiotensin system. Renin is released from cells of the kidney in response to many stimuli, including low sodium concentrations and low blood volume. This then helps the bird return to normal levels of sodium and blood volume.

Adrenal Medullary Secretions

The medullary secretions are epinephrine and norepinephrine (adrenaline and noradrenaline, respectively) and are also called catecholamines. These are both mainly neurotransmitters of the adrenergic nervous system. They exert profound effects on the cardiovascular system and carbohydrate and lipid metabolism. Glycogenolysis is stimulated by adrenaline, and lipogenesis is inhibited by adrenaline and noradrenaline. Lipolysis is stimulated in the chicken by adrenaline.

Thyroid Glands

The thyroid glands in birds are paired organs oval in shape and dark red in color. They are located on either side of the trachea on the ventral-lateral aspect of the neck just exterior to the thoracic cavity and can be found adhering to the common carotid artery just above the junction of the common carotid with the subclavian artery. At day 7 of incubation,

---

Temperature Regulation

Birds are *homeotherms,* meaning they maintain a relatively constant body temperature. They are also *endotherms,* meaning that they can also help maintain their body temperature or increase it by generating heat within their tissues. Birds and mammals have many features of thermoregulation in common; however, the differences are striking.

Feathers serve for flight and insulation in birds and the disposition of fat tends to be different in birds and mammals and is related to insulation. Birds do not have sweat glands but are very efficient at respiratory evaporative cooling.

Deep-body temperature may differ from skin or peripheral temperature. The cardiovascular system plays a major role in maintaining differential temperatures and aiding in the removal of excess heat from the core to be lost at the surface. The average deep-body temperature is about 106.7°F (41°C). This temperature fluctuates with time of day and activity and follows a circadian rhythm.

The range of environmental temperatures at which birds exchange heat with the environment the least is known as the *zone of least thermoregulatory effort* or the *zone of thermal comfort.* This has traditionally been called the *zone of thermoneutrality.* This means heat is neither lost nor gained by the animal, but this rarely occurs. The typical range is 71.6 to 78.8°F (22–26°C).

Body temperature is determined by the balance between the amount of heat produced by the animal, gained from the environment and lost to the environment. The concept is illustrated by the heat-balance equation:

the thyroid of the embryonic chick can concentrate radioactive iodine. It becomes functional and secretes T_4 by 10 to 11 days of incubation.

The thyroid is capable of concentrating iodine when the blood levels are extremely low and can trap iodine about two to four times the rate of mammals. Both T_3 and T_4 have equal potencies for preventing goiter, stimulating body weight and comb growth, and influencing oxygen consumption and heart rate.

$$M = E \pm R \pm C \pm K \pm S$$

where M = metabolic heat production;
E = evaporative heat loss (positive);
R = heat loss by radiation (positive if heat is lost to the environment);
C = heat loss by convection (positive if heat is lost to the environment);
K = heat loss by conduction (positive if heat is lost to the environment); and
S = heat storage (positive if heat content of the body increases and body temperature rises).

The comb, wattles, shanks, and feet are exposed surfaces of the body where increased blood flow will result in increased heat loss to the environment. Birds standing on cool roosts (water pipes with cool water flowing in them) will maintain normal body temperature during periods of heat better than birds on standard roosts.

Temperature regulation is under the influence of the hypothalamus and the spinal cord.

Much of the heat loss of birds is due to thermal panting and respiratory evaporative cooling, especially as the temperature increases past 84.2°F (29°C).

Preventing heat loss to the environment is also accomplished by reducing blood flow to the surfaces and by countercurrent exchange between blood vessels in the limbs, especially the shanks.

Birds allowed free access to their environment also rely on behavioral thermoregulation. This means they will seek out the most energy-efficient means of maintaining their body temperature, such as ruffling their feathers, drinking water, moving into the shade or sun (or heat source), huddling, and lying down.

The function of the thyroid is governed, within certain limits, by the circulating levels of thyroid hormones and feedback on the hypothalamus to stimulate release of TSH from the pituitary. Some studies have shown that testosterone increases peripheral utilization of thyroid hormones.

Parathyroid Glands

The ability of birds to produce an egg each day for extended periods represents a calcium demand of approximately 10 percent of the total body stores. Thus, chickens must possess an efficient and effective mechanism for maintaining calcium homeostasis. The parathyroids have been implicated in this phenomenon.

The four parathyroid glands are located slightly to the rear of the thyroids, a pair of glands found on each side of the midline. Chicken parathyroid hormone (PTH) has a primary target of bone and kidney. Injection of PTH results in a rapid rise in blood calcium levels. Parathyroid has been implicated in the renal synthesis of the hormonal form of vitamin D_3; it is also believed that PTH affects resorption of calcium from medullary bone.

Ultimobranchial Glands

The ultimobranchial glands of the chicken are located posterior to the parathyroid glands on either side. They secrete the hormone calcitonin. Calcitonin levels increase in the plasma in relation to increased plasma calcium levels. They appear to maintain the balance between ionic (free or usable) and bound calcium in the blood. The full role of calcitonin in chickens has yet to be elucidated.

Pineal Gland

The pineal is located between the cerebral hemispheres and the cerebellum of the brain. Pinealocytes can actively take the amino acid tryptophan from the blood and transform it into 5-hydroxytryptamine (5-HT) and also produce the hormone melatonin. The peak of melatonin rhythm has been shown to occur in midscotophase (complete dark). Melatonin has been shown to affect sleep patterns, behavior, as well as brain electrical activity.

Some evidence shows that melatonin exerts significant influence over pituitary gonadotropins by affecting the hypothalamic releasing factors for LH and FSH. In chickens, it has been shown that melatonin has an inhibiting influence on LH release and that it inhibits ovulation and decreases the weight of the testes and the ovaries of developing chickens. This may account for the fact that birds do not produce well on short days and long nights. The pineal also plays a role in the maintenance of body temperature, weight, and adrenal size.

Immune System

The immune system of chickens is slightly different from that of mammals in that birds have a *bursa of Fabricius*, which produces B lymphocytes. Other organs involved with immunity are the thymus, harderian gland, cecal tonsil and Peyer's patches, spleen, lymph nodes, and pineal gland. Immunity is subdivided into humoral and cell-mediated. Humoral immunity is provided through the T and B lymphocytes, which produce immunoglobulins, such as immunoglobulins A and G (IgA and IgG), and antibodies. Cell-mediated immunity includes immunity not dependent upon the synthesis of antibodies. Examples of cell-mediated immunity are graft versus host (such as transplants of organs or skin tissue).

The main function of the immune system is to destroy foreign invaders in the body, such as bacteria, viruses, and protozoa. Some immunity is passed from the hen to its offspring through immunoglobulins and antibodies in the egg yolk and albumen. Vaccines provide immunity for many diseases of poultry. Vaccination schedules vary with the breed and species. Many vaccines can now be given in ovo, into late-stage embryos. Proper nutrition and environment can help maintain a healthy immune system.

Four

STARTING THE LAYING FLOCK

You can start the laying flock in a variety of ways: Set fertile eggs under a broody hen or in a small incubator; purchase day-old chicks or started pullets; or begin with second-year layers. (These are birds that have laid for 1 year.) Production is not as good for second-year layers; egg quality may be poor if they have not been molted; or soon after moving they may stop laying and go into a molt (that is, shed their feathers and grow new ones).

One of the simplest and best ways of starting the laying flock is to buy started pullets. This avoids the need for equipment and the care involved in incubating eggs and brooding chicks. You can buy started pullets at 15 to 20 weeks of age from commercial growers. Sometimes it is possible to buy a few surplus pullets from a local poultry producer. Pullets should be beak trimmed to avoid feather picking and cannibalism. Five to ten birds will normally provide the average family with enough eggs. Unless surplus eggs can be marketed easily, you should not produce more than the family needs. Bear in mind that a healthy, well-managed laying flock can produce 21 to 22 dozen eggs per bird in a laying year.

Choosing the Best Chicks

Before buying birds, decide whether you want white or brown eggs. Eggshell color does not affect nutritional value, but it does influence the market price, because there is a definite preference in various areas of the country. Most sections of the United States are predominantly white-egg markets, but in New England brown eggs are preferred. Brown eggs usually command a higher price.

Birds that lay white eggs typically weigh in the vicinity of 3½ to 4½ pounds (1.6–2 kg) at maturity. Representatives of these breeds and varieties are the Leghorns, first-generation Leghorn-strain crosses, and hybrids — all of which are good egg producers but not so good for meat.

The medium-weight American breeds lay brown eggs. The mature weight of these birds is usually between 5 and 6 pounds (2.3–2.7 kg), thus making them a little better for use as meat birds at the end of their laying period. Most of the medium-weight, brown-egg-producing layers are crosses of American breeds, such as the Barred Plymouth Rock and the Rhode Island Red, and are generally known as sex-links. There are several of these crosses, all of which are good egg producers. Several pure breeds, such as the Rhode Island Reds, White Plymouth Rocks, and New Hampshires, can be used as dual-purpose breeds; that is, for meat and eggs. However, they may not do as well as the crosses bred especially for meat or egg production.

Stock bred specifically for high egg production or for meat production is best suited to the home flock. To get the project off to a good start, purchase well-bred, healthy chicks or started pullets. Purchasing chicks from distant hatcheries may expose them to less than desirable conditions when being transported over long distances. They may be chilled or overheated, leading to early losses and a poor-producing flock. It is usually best to purchase chicks or started pullets from a reputable nearby producer or hatchery that has stock bred for efficient production.

Traits to Consider

Consider the following traits when purchasing chicks or pullets: livability, early feathering, feed efficiency, freedom from disease, rate of growth, egg color, egg size, egg production, and egg quality. You can often obtain much valuable information by talking with poultry producers in your area. They can usually tell you from experience which breeds and strains have desirable characteristics and the most reputable hatcheries and pullet growers from which stock can be obtained.

One of the first steps toward getting disease-free chicks is to buy the stock from hatcheries that blood-test their breeders for *Salmonella pullorum*, *Salmonella enteritidis*, and typhoid. Hatcheries operating under the National Poultry Improvement Plan blood-test their breeding flocks for these diseases, which are passed from the infected hen to the chick through the egg.

If a nearby hatchery has stock with good production capabilities and is disease-free, it is best to get your chicks locally to minimize the shipping time and avoid the possibility of prolonged exposure to extremes of temperature and poor handling. It is usually much easier to get adjustments from a nearby hatchery in the event that problems arise. Chicks should be ordered at least 4 weeks in advance of the date you would like to start them, and started pullets should probably be ordered close to 6 months before they are needed. Your county Extension agent or Extension Poultry Specialist can help you select sources of stock (see the directory of Cooperative Extension System Offices on page 312).

Additional Information

For a list of hatcheries participating in the National Poultry Improvement Plan, write to USDA, APHIS — VS, National Poultry Improvement Plan, Suite 200, 1498 Klondike Road, Conyers, GA 30094.

Buying Your Chicks

The importance of selecting the right stock cannot be overemphasized. After deciding on the breed or strain you want, be sure to buy healthy stock from hatcheries with a U.S. *Salmonella enteritidis–*, *Salmonella pullorum–*, and typhoid-clean status. Purchase the best chicks you can afford. Cheap chicks may cost you more in the end than those that cost most in the beginning. Poor results with the growth or livability can soon erase any savings you may have made in the purchase of chicks.

Be sure to buy chicks bred for the purpose for which they are to be used. If you are interested in starting with a flock of laying birds, get chicks from a source known for having birds with high-egg-production records. There are great differences among types of birds, insofar as egg size, egg production, feed efficiency, liability, and many other economic traits are concerned.

In the New England area, production birds that lay brown eggs are primarily the sex-link crosses. Several crosses of this type are offered for sale, known by various names. They are called *sex-link crosses* because the breeds that are mated result in male chicks of one color and female chicks of another. Day-old chicks can be sexed by feather color or primary flight-feather length, thus eliminating the cost of vent sexing.

When buying chicks for laying stock, the usual practice is to purchase sexed pullets. This is because the cockerels of laying stocks are poor meat producers; and it is not wise to try to grow them for meat.

Only birds bred for meat production should be raised as broilers, capons, or roasters. Stock that is bred for the production of meat has the ability to grow rapidly with a minimum of feed consumption. Feed consumption represents at least 60 percent of the total cost of producing poultry meat; thus, feed cost becomes an important consideration. Also, birds bred for meat grow faster with a better finish, and are more tender and juicier. The meat-bird stocks are primarily white feathered, which results in a much better-dressed appearance in the absence of dark pinfeathers.

Sex-link layer (popular brown-egg New England crossbred)

When to Start Chicks

The time to start chicks depends upon the type of housing you have and the type of brooding equipment available, as well as availability of labor, time, and money. Chicks started in the winter commence laying in the early summer, but may go into a neck molt and take a vacation the following winter.

Probably the best time for most small-flock producers to buy their chicks is in late March, April, and May, especially for those in the northern climates. Chicks started at that time do not need cold-weather brooding facilities. Moreover, the pullets will start laying in early fall and will continue laying for a full cycle if properly managed.

Breeds of Poultry

Only a few of the several hundred breeds of poultry will be described here. For more information on these and other breeds, the reader is directed to the *American Standard of Perfection*, published by the American Poultry Association.

Chickens are listed in several classes: American, Asiatic, English, Mediterranean, Continental, Bantams, and Games. The most popular type of chicken for commercial egg production is the White Leghorn. Originally from Italy, it spawned many subvarieties in America, Denmark, and England. They are noted for their prolific egg production and white-shelled eggs. Leghorns are lightweight birds and not good setters.

Rhode Island Reds

The Rhode Island Reds were developed in 1904 from a cross of Red Malay game, Leghorn, and Asiatic stock. They are known as dual-purpose birds because of their high egg production and their potential for meat production. They are a popular brown-egg producer for small poultry flocks. Although not used as much for meat production, the hens can attain body weights of 5½ to 6½ pounds (2.5–2.9 kg) and the males from 7 to 8 pounds (3.2–3.6 kg). They are also a popular show breed. Rhode Island Reds are a deep dark-red color.

Rhode Island Red

New Hampshires

New Hampshires were developed from the Rhode Island Reds. They were selectively bred for early maturity, large brown eggs, and fast feathering. Special meat strains have been developed for fast growth and increased weights, but do not have the egg-laying quali-ties of the dual-purpose Standard New Hampshires. The New Hampshires are a chest-nut red, lighter in color than are the Rhode Island Reds.

New Hampshire

Barred Plymouth Rocks

Barred Plymouth Rocks were originally bred from composites of several bloodlines. The most prominent cross was that of a Dominique male with Black Cochin or Black Java females. Although the Barred Plymouth Rock is well known, White, Buff, Silver Penciled, Partridge, Blue, and Columbian Plymouth Rocks are also in the class of dual-purpose birds. The females weigh from 6 to 7½ pounds (2.7–3.4 kg) and the males, 8 to 9½ pounds (3.6–4.3 kg). They have yellow skin and lay brown eggs. Barred Rocks are a popular backyard bird.

Barred Plymouth Rock

Wyandottes

Wyandottes are another breed of dual-purpose brown-egg layers. Although Silver Laced Wyandottes were the parent strain, Golden Laced, Buff, Partridge, Black, Columbian, and White Wyandottes are all standard varieties available today. Hens weigh in at about 5½ to 6½ pounds (2.5–2.9 kg), while males range from 7½ to 8½ pounds (3.4–3.9 kg). They are a hearty breed and popular with backyard and small-flock owners.

Wyandotte

Australorps

Australorps are production-bred Australian Black Orpingtons and are noted for their high egg production, but are still considered one of the dual-purpose breeds. Their all-black feather color makes them a striking bird for the home production flock, as well as for show.

Australorp

Araucanas and Ameraucanas

Araucanas and Ameraucanas are sought after for their turquoise- or blue-shelled eggs. Little is known of the Araucanas except that they came from South America. They were first imported into the United States in the early 1930s. They are rumpless chickens and have tufts of feathers protruding from each side of their neck. The blue-egg-color trait is dominant and will occur in offspring when crossed with other breeds, and these crossbreds with blue eggs are sometimes mistaken for Araucanas. Ameraucanas were developed in the United States in the 1970s to eliminate some of the undesirable characteristics of the Araucanas, such as temperament and lack of rump. The Ameraucana has a muff, a tail, and no ear tufts and comes in several varieties, such as Black, Blue, Buff, Silver, and Wheaten.

Araucana

Cornish

Cornish fowl originated in Cornwall, England, as a composite of several bloodlines. The Cornish are one of the foundation stocks of modern meat chickens, along with the White Rock and other heavy breeds. They are distinguished by their large, heavy breast and thick, stocky legs.

Cornish

Ornamental and Game Breeds

There are several breeds whose main qualities are for ornamental or show purposes. Among these are the Crested Polish, Houdans, Faverolles, Brahmas, Cochins, Langshans, Phoenix, and Yokohamas. Other breeds are selected as game birds, such as the Old English and Modern games.

Many of the standard breeds are also available as bantams, which are about one-quarter to one-fifth the weight of their standard counterparts. Bantams are very popular as pets and for show.

Other breeds of ducks and turkeys are described in chapters dealing with these types of birds.

LIVING LAWN ORNAMENTS

Mary K. has a small flock of fifteen Barred Plymouth Rocks. She lives in a semirural area on 2½ acres and says her birds are her pets and "living lawn ornaments" for her yard. She fondly recalls that her grandmother had chickens around her home and how much she loved the fresh eggs.

Mary has one rooster, Mack, who keeps watch over the ladies (hens). She has a nice 8-foot-by-10-foot coop the birds stay in at night and where she provides feed and water. She has a roosting area made from tree branches about 1½ inches in diameter, and seven nest boxes. During the day the birds roam around the yard, getting into everything. Mary also has a chocolate Lab that plays with the chickens and even allows Mack to sit on its back.

She sells her extra eggs to friends and neighbors to help cover the cost of feed. When her grandchildren come to visit, the birds go right up and greet them and allow the children to pet and touch them and feed them table scraps.

For Mary and her family, poultry have become a way to learn about and appreciate the diversity of animal life. They especially love to sit on the porch on a Sunday afternoon and watch the birds as they wander around the yard, pecking and scratching at the ground and catching bugs. Mary, like many other small-flock owners, has found that just sitting back and observing her birds after work is a great form of relaxation therapy, helping keep her a calm, contented person.

FIVE

BROODING AND REARING YOUNG STOCK

Brooding facilities can affect the degree of success one has in growing chicks, poults, ducklings, and goslings. As indicated earlier, the building need not be expensive, but it should be built to brood chicks efficiently.

One of the features of a good brooder house is structural soundness to withstand high winds and snow loads in cold-winter areas. It should be of tight construction to avoid drafts and leaking rain, with sufficient insulation, making it easy to maintain a uniform temperature economically. It should have a ventilation system that controls moisture and gases, yet still is able to conserve or dissipate heat. Last, it is desirable to have some degree of light control in the house.

Equipment

The main equipment needs are feeders, waterers, and brooders. Roosts may be used for replacement pullets. (See chapter 2 for details.)

Feeders

For the first couple of days, it is best to feed the chicks with a large flat container with an edge of about 1 to 2 inches (2.5–5 cm) in height. You can do this with chick-box lids (without holes), commercial feeder trays, or egg flats filled with feed. After a couple of days, you can feed them using commercial or homemade feed troughs. The commercial troughs may be easier to clean and result in less feed wastage. Round hanging feeders (tube feeders) are excellent for small flocks.

They are easily adjusted to accommodate birds of different ages and to avoid feed waste. Because they have storage space, you won't have to refill them as frequently as the trough type, and the feed stays cleaner.

Waterers

It is possible to make water fountains using large juice or fruit cans in combination with tin plates. Make two holes on opposite sides of the can about ¾ inch (1.9 cm) from the lip. Fill the can with water, place the plate on top, and flip the can and plate over to create a waterer that maintains its own water level. See page 29 for illustrations of this waterer.

To minimize waste, feeding equipment should be designed so that it can be adjusted as birds grow. Keep the lip of the hopper at about the level of the birds' back.

A small wire platform underneath each fountain helps keep the water clean and the shavings out of it, and also prevents the litter around it from becoming too damp.

Probably the best watering device for chicks is the glass- or plastic-jar 1-gallon (3.8 L) fountain. Commercial poultry producers frequently use the gallon fountains in combination with automatic waterers. They gradually move the gallon fountains close to the automatics. They are thus able to convert to the automatic waterers as early as 10 days of age. Once the chicks are drinking well from the automatic waterers, the fountains can be removed gradually and additional automatic waterers added until they have three or four 8-foot (2.4 m) waterers per 1,000 birds.

In small-flock situations, two 1-gallon (3.8 L) waterers are usually good for one hundred birds for up to 2 weeks. From weeks 2 through 6, there should be at least two 2- or 3-gallon (7.6 or 11.4 L) waterers, and from 8 to 10 weeks, one 5-gallon (18.9 L) waterer per one hundred birds.

Brooders

To provide the necessary heat, you will need some type of brooding device. Infrared lamps are an excellent source of heat for brooding small flocks of chicks (see page 31). The initial cost of equipment is relatively small, and the lamps are excellent for supplying heat to the chicks, particularly in the late spring, summer, or early fall. Suspend the infrared lamp approximately 18 inches (45.6 cm) above the litter. One 250-watt bulb is sufficient to brood as many as seventy-five chicks. Even though the number of chicks may be small, it is wise to use at least two bulbs in case one fails.

Many small flocks are brooded with homemade brooders utilizing electric lightbulbs as the heat source. It is possible to construct simple hovers (canopies) of plywood and canvas, or some similar material, which you can suspend from the ceiling or support with legs (see page 31).

Other brooders, manufactured commercially, are fired by oil, gas, electricity, wood, or coal. These have a hover to keep the heat at floor level (see page 31). Such brooders are excellent for flocks of one hundred chicks or more. Some of them are rated at a capacity in excess of 750 chicks. The initial cost of these units is considerably more than that of either the infrared lamp or the homemade brooder. Each chick needs a minimum of 7 square inches (45.2 sq cm) of hover space or its equivalent.

Each type of brooder has certain advantages. The infrared bulb has the advantage of enabling the operator to see the chicks at all times. Where the hover type of brooder is used, it is more difficult to observe the chicks. On the other hand, the hover brooder is best for cold-weather brooding and is less expensive to operate.

The type of equipment you select will depend somewhat upon the cost and availability of fuel and the extent to which it is to be used. Whatever brooder stove you use, it should have the capacity to heat adequately and a means for controlling the temperature accurately.

Roosts

As noted in chapter 2, roosts should *not* be used for broilers, roasters, or capons. In fact, few people use roosts even for growing replacements. Those who want their birds to roost in the laying house feel that it is necessary to train the pullets to roost in the brooder house. If you use roost perches, enclose them with poultry netting to prevent the chicks from coming into contact with the droppings. Construct roost perches to provide easy access for the birds — often they are slanted from the floor to the wall much like a ladder.

Preparation of the Brooding Quarters

Getting the chicks off to a good start is very important. Probably no other part of the enterprise deserves more planning and preparation than does the starting of baby chicks.

1. Clean and disinfect or fumigate the brooder house at least 2 weeks before the chicks arrive. If you do not do this, the chicks may be exposed to certain poultry diseases that can lead to mortality and poor results. First, dry-clean to remove the dirt, dust, and debris from your building. Then wash down the facility using a good detergent and rinse thoroughly. After the building is clean, use a good disinfectant. Disinfectants are effective only on clean surfaces. There are a number of good commercial disinfectants available containing active ingredients such as quaternary ammonium compounds, phenolics, peroxides, and chlorine. Use them according to the manufacturer's recommendations. Some of them require that the house stand idle for a period of time and be thoroughly aired before the chicks arrive. Failure to observe these precautions may cause the chicks to have severely burned feet and eyes.

After you have cleaned and disinfected the brooder house, allow it to dry thoroughly before putting in new litter.

2. Prepare the litter. When the floor is dry, cover it with 2 to 4 inches (5.1–10.2 cm) of litter material. The litter material serves to absorb moisture and to insulate the floor for the birds' comfort. Wood shavings, when available, are among the most commonly used materials for the litter. Others include sawdust, or shavings and sawdust, rice hulls, sugarcane, peanut hulls, and ground corncobs.

Important Note

Chicks that are very hungry upon arrival may try to eat litter before learning to eat feed. To prevent this, put paper over the litter for the first 3 or 4 days after they arrive until they learn to eat from feeders. This is not necessary for chicks received within a day of hatching.

---

3. **Start the brooders a day or two before the chicks are due to arrive.** This will ensure that the equipment is operating properly and is adjusted to the correct temperature. Use a chick guard (brooder guard) to confine the chicks to the source of heat. The chick guard keeps them confined to a given brooder and also prevents migration and over-crowding of some brooders if more than one is used in the facility. It will also help prevent drafts on the chicks under the brooder. A corrugated-cardboard guard, approximately 12 inches (30.5 cm) high, is good for this purpose during cool weather. For warm-weather brooding, the guard may be made of poultry netting. It should form a circle around, and about 3 feet (91.5 cm) from, the brooder. You can extend the brooder guard out-ward after a few days. It is usually removed at day 10 or so, depending upon the weather and conditions in the brooder house.

4. **Fill the feeders and waterers several hours before the chicks arrive.** The water will then be at room temperature, and you should encourage them to drink. Space the feed trays and water fountains uni-formly around the brooder and close to the hover. Chicks instinctively peck at anything at the same level as the surface upon which they are standing. Therefore, they will learn to eat more readily if you provide the feed in feed trays or egg flats for a day or two. Chicks that are slow to catch on to the fact that feed is in front of them are attracted by the noise made by the others that have found the feed and are pecking on the card-board or plastic. This gets the chicks off to a good start, eating well.

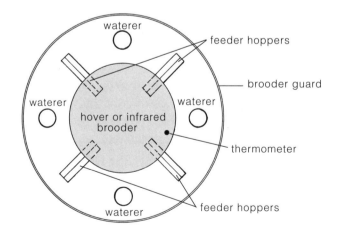

Location of trays and waterers around the brooder

Brooder Management

Brooding temperature and good litter conditions are essential for successful brooding.

brooder house drafty

brooder just right — chicks comfortable

brooder too cold

brooder too hot

Brooder management

Brooding Temperature

Brooding temperature is very important. The recommended temperature at the start is 90 to 95°F (32–35°C). A rule of thumb to remember: Reduce the temperature 5 degrees each week until the chicks no longer need heat. It is often stated that the best thermometer, the way to gauge the most comfortable temperature, is to watch the chicks themselves. After the chicks reach 7 days of age, this is probably true. If they huddle close to the heat source, you can be reasonably sure that the operating temperature is too low. If, on the other hand, the birds are located in a circle, way outside the heat source, you should assume that the temperature is too high. During the day, the chicks should be evenly distributed around the entire brooding area, with some of them underneath the heat source. Take temperature readings at the outside edge of the brooder at chick level.

Litter Conditions

One important factor for the successful brooding of chicks is the maintenance of good litter conditions. When they get out of hand and the litter becomes too wet, disease can result. Wet, dirty litter can harbor many disease organisms that affect poultry. In the case of replacement pullets, it is desirable to have a certain amount of moisture in the litter (30 to 35 percent) to enable sporulation of oocysts and the subsequent development of an immunity to coccidiosis during the growing period. However, an excessively wet litter, especially during warm periods, can bring about a clinical case of coccidiosis requiring treatment.

A common problem experienced with meat birds is breast blisters — swellings or external sores on the skin of the breast — which seriously detract from the dressed appearance of the carcass. These are quite common in flocks of capons and roasters. Excessively wet, caked, or dirty litter is frequently blamed for breast blisters. Since perches are not used for broilers or meat birds, the birds must bed down in the litter. It is important that the litter depth be maintained and that it is kept dry and fluffy enough to cushion the body and avoid all contact with the floor. Furthermore, it is essential that it be reasonably clean to avoid soiling of the breast feathers, skin irritations, and breast blisters. If you pick up a handful of litter and squeeze it and water comes out, it is too wet; if it just crumbles, it is too dry; if it packs into a ball and holds its shape, it is just right. Aerating the litter by stirring and fluffing will help keep it in good condition.

General Feeding Programs

The recommended modern feeding programs for meat birds and laying-bird replacements vary widely. Feed broilers and other meat birds a high-protein, high-energy diet throughout the growing period. In the production of meat birds, the goals are rapid growth, heavy weight, and efficient feed conversion. Start replacement pullets, on the other hand, on medium- to high-protein and energy starter rations for the first 6 to 8 weeks; then change them to a grower diet to accommodate their changing dietary needs.

Meat-bird chicks and replacement pullets reared on litter are normally started on a diet containing a coccidiostat. Use a higher level of coccidiostat for meat birds to prevent coccidiosis, a protozoal disease common to chickens reared on the floor. The coccidiostat is usually fed throughout the growing period to meat birds to protect them against a coccidiosis outbreak and to obtain maximum growth and efficiency. Feed replacement pullets a lower level of coccidiostat in the starter diet. This enables them to develop an immunity to coccidiosis by 6 to 8 weeks of age, and this immunity protects them throughout the remainder of the growing and laying periods.

Diets for meat birds, particularly broilers and turkeys, are usually somewhat better fortified with vitamins and growth-promoting factors. They sometimes contain very low levels of antibiotics to help maintain the proper balance of gut microorganisms and to enhance feed efficiency and growth. The protein level of meat-bird diets, at least during the first several weeks, is higher than for replacement birds. Also, feeding programs for meat birds do not include scratch grains during the growing period. Occasionally, heavy meat birds are fed supplemental corn the last 2 weeks of the growing period to improve their finish just prior to dressing or marketing. A special finishing diet is often used for this purpose. Some growers of meat birds prefer the crumbles or pellets to the mash form because of less feed waste and increased feed consumption.

Some custom-blended broiler or roaster feeds contain feed additives that must be removed several days prior to slaughter. The withdrawal period varies with the type of drug — for the last several days, feed the birds on a special feed without additives. Some drugs used for the treatment of diseases are administered through the feed. Definite withdrawal periods are required for these drugs prior to slaughter. Always follow the recommendations of the feed company printed on the feed tag.

Safe Feeds?

With the increased interest in *natural* or *organically grown* poultry, attention has been focused on how the birds are fed and reared. Unfortunately, many people don't understand the feed and rearing practices for poultry, which sometimes causes wariness of poultry products in general. Let us clear the air.

Synthetic estrogens were once used instead of surgery to caponize chickens; however, this practice was banned in the United States almost 40 years ago. Some commercial broiler growers supplement their poultry feed with subtherapeutic levels of specially approved antibiotics to improve the growth of their birds. This is accomplished by reducing or preventing digestive upsets and diarrhea by keeping the gut microorganisms in proper balance. However, unless specifically ordered in the feed, *no poultry feeds contain antibiotics.* Antibiotics cannot be added to feeds available to the small producer unless they are ordered by a veterinarian. Reputable, brand-name feed manufacturers must use the highest quality ingredients for their poultry feeds, or they risk losing their customer base.

Feeding and Management of Layer Replacements

Replacement chicks for layers are started on a diet similar to the broiler diet; however, the energy and protein contents are usually slightly lower. The protein content is 20 to 21 percent. Chicks to be raised for layers are normally fed a starter for 6 to 8 weeks and then are changed to a grower diet.

As with other animals, the growth rate diminishes as the birds become older. Therefore, the protein requirement is less later in the growing period, and the grower diet is formulated with somewhat less protein than the starter.

Several feeding programs are used from the 6- to 8-week period through housing. The two basic systems are the all-mash diet with approximately 14 to 16 percent protein or a mash and scratch diet.

Why Delay Sexual Maturity?

Birds that commence to lay too early or that are too fat at the onset of production frequently do not perform as well as slower-maturing birds. They are inclined to lay fewer and smaller eggs, and are more prone to prolapse of the uterus and higher mortality.

Birds on a full-feed program that are grown in the off-season (under conditions of increasing day length) should receive a controlled lighting schedule to delay sexual maturity. This method, if followed correctly, usually produces the most economical gains and more uniform birds with fewer social problems or vices.

One feeding program still used fairly commonly is a combination of the grower concentrate with a grain such as oats. After 8 weeks, the oats can be gradually added to the grower diet until proportions of approximately one-half grower concentrate and one-half oats is reached. Some producers permit the birds to free-choice feed the mash and oats. Use a hard, insoluble grit, such as calcite crystals, when a whole grain is fed. Use a medium or pullet-sized grit to approximately 12 weeks of age and the hen-sized grit, or coarse, for the remainder of the growing period. Grit must be present in the gizzard to enable birds to grind the fibrous material so it can be utilized and absorbed into the intestines.

Range Rearing

Under some conditions, range rearing, or yarding, may be recommended. With all the interest by the public in organic or free-range eggs and chickens, range rearing has gained in popularity. It may be possible to save some of the cost of feed and still grow good healthy pullets. Practically all broilers and most other meat birds are reared in confinement, so as to have absolute control of the feed intake, more rapid and efficient growth, and better finish. As mentioned earlier, rapid growth and early maturity are not desired in replacement birds, so range rearing or yarding may be used for replacement pullets.

Range rearing was once a popular method of growing young stock during warm weather. There is no question that excellent pullets can be grown on range. Good range-reared birds usually have excellent shank color, beautiful feathering, and are inclined to be big-framed birds with health and vigor. Why, then, have many producers gone to confinement rearing?

Drawbacks

Losses from predators and the increased labor requirements in feeding and caring for the birds are probably two of the biggest reasons for getting away from range rearing of commercial pullets. Predators such as foxes, raccoons, coyotes, and even hawks and owls can be a real problem when birds are kept on range. Then there is the expense of building fences and range shelters, and their upkeep. If the birds are to be yarded, you can perhaps utilize the brooder house for shelter, thus avoiding the need for special range shelters. The choice of whether to range or confine birds will depend upon a number of factors, such as the size of flock, the labor situation, and the land and housing facilities available. The shelter on page 14 is suitable for about one hundred pullets.

Feed

If you manage your range properly, it will yield a good supply of forage for growing pullets. Some of the most satisfactory grasses and legumes for poultry range include ladino clover, Kentucky bluegrass, and bromegrass. This type of range permits feeding of a grower diet on a restricted basis and makes the birds forage for a portion of their nutrient intake. Keep the range clipped to be most palatable and in a sanitary condition — that is, without bare spots and mud holes.

ladino clover

Kentucky bluegrass

bromegrass

Facilities

Move range shelters and equipment occasionally when the ground around them becomes bare to keep the birds out of the mud. It is desirable to change range areas, so that a given area is not used every year. This may help avoid disease. Capillaria worms are one of the problems experienced on wet, muddy range.

Covered feeders are recommended when birds are fed on range. They will help prevent feed loss or spoilage from wind and rain. A covered range feeder is shown on page 27.

It requires an acre (0.4 ha) of good range to properly handle four hundred to five hundred pullets. Range rearing is limited to the warm seasons of the year or to areas with warm climates.

Small flocks of layers, pullets, roasters, or capons are frequently provided a yard. The yard is a fenced-in area attached to the brooder house or some other shelter. A yard permits more birds to be kept in a given floor area and may help prevent management problems such as cannibalism and feather picking. Its use in the cold climates is pretty much restricted to the warm months.

If you use a yard, the feeding and watering equipment is usually left in the house. If you use covered feeders, you may do some of the feeding in the yard.

The same hazards may exist with yards as with range or pasture. Disease may be a consideration if you use the same yard year after year. Ideally, you should move the yard area each year. Predators may also be a problem. Layers permitted to yard may hide their eggs in areas outside the house. In wet weather, the eggs may be excessively dirty. It is usually best to confine the laying flock for optimum results.

Vaccination

Several vaccines are available to prevent outbreaks of a number of the troublesome poultry diseases. Some of the common immunizations are for infectious bronchitis, Newcastle disease, laryngotracheitis, fowl pox, and cholera. In some areas, all of these diseases are a problem. Newcastle and bronchitis can be problems anywhere, and fowl pox is prevalent in many areas. In northern New England, we don't see fowl pox very often and do not recommend vaccination for it. The same has been true for laryngotracheitis. Vaccination programs should be established on the basis of those diseases prevalent in the particular area. Commercial producers routinely vaccinate for Newcastle and bronchitis in all areas of the United States and, in some areas, also vaccinate for fowl pox and cholera on a routine basis. Vaccination recommendations in various areas, therefore, will differ, and your program should be developed to handle the disease situation in your area.

Some of these vaccinations come early in the chick's life; the Marek's vaccination, for example, is administered at 1 day of age, usually at the hatchery. Many large commercial hatcheries are now vaccinating the eggs a few days prior to hatching. Incidentally, it is advisable to purchase chicks that have been vaccinated for Marek's disease. For many years, the disease took a tremendous toll on growing birds. With the advent of the Marek's vaccine, mortality has been cut to a negligible rate. Other vaccinations are administered at various times during the growing period and may require booster shots during the laying period.

Types of Vaccines

There are two basic types of vaccines: those that produce a temporary immunity and those that produce a solid or permanent immunity, which will protect the birds for the remainder of their lives. The virus vaccines producing a temporary immunity are usually of low virulence or, in some cases, have no apparent effect on the birds when administered. The virus vaccines producing a permanent immunity are of a more severe virulence, and therefore will produce more marked or serious symptoms of sickness.

Application

Vaccines are applied to individual birds or to the flock on a mass basis. Vaccines for Newcastle and infectious bronchitis are applied on a mass basis as a coarse spray, a fine mist, or in the drinking water. Those vaccines used on a mass basis are mild vaccines of low virulence.

Some vaccines are not permitted in certain states. Vaccination of small flocks is not recommended in many areas. Contact your poultry diagnostic laboratory or the poultry specialist at your state university for more information. Local industry service people and county Extension agents can also be helpful.

Plan Your Vaccination Program Carefully

Because there are so many factors to consider, and so many variable conditions from farm to farm, no set vaccination program can be recommended that will best suit all farms. Your program should be planned to meet your needs. It should be planned well in advance of the brooding and rearing season, and be scheduled so that all the birds that are to be placed in laying houses will have been vaccinated well before they are ready to lay. This will give them sufficient time to fully recover from the stress of vaccination and avoid any carryover effects on egg production or egg quality.

Feather Picking and Cannibalism

Both feather picking and cannibalism are rather common vices that can develop in the brooder house and may be carried on into the laying house. A similar type of problem is toe picking, which can start in a flock of chicks soon after they are put down under the brooder.

The exact cause of these vices is not always evident. Poor nutrition; overcrowding; overheating the birds; lack of floor, feed, and water space; and even very bright lighting are all thought to be factors. Boredom and increased daylight hours may also contribute. Sometimes the problems occur even with apparently good management.

At the first sign of feather picking, take action to stop it immediately. When blood is started, the birds frequently pick each other apart and losses can be severe. In some flocks, the problem can be corrected by a change in management — providing more feed, water, or floor space, better ventilation, cutting back on the light, or any number of things that will improve the birds' comfort or positively change environmental conditions.

In the case of small flocks, antipick ointments applied to the birds have been somewhat successful in curtailing the problem. Plastic and metal devices known as *specs* can also be used. Specs are mounted on the top beak and are designed to obstruct the bird's forward vision. These devices can catch on the poultry furniture and come off.

For some flocks, changing from crumbles or pellets to a mash feed will result in more time spent at the feeder, with less time for fighting.

Sometimes the additions of environmental enrichments will help reduce the effects of cannibalism. Hanging a head of cabbage by a string at the height of the birds' heads provides a good diversion. The birds pick the cabbage leaves and chase each other around trying to get it.

Beak Trimming

The most widely used, and probably the most satisfactory, method of cannibalism control is to beak trim the birds. Many producers beak trim their birds routinely to prevent picking problems.

Chicks can be beak trimmed at the hatchery at 1 day old, using a precision beak-trimming machine. If done properly, beak trimming done at 6 to 8 days of age is very effective and most beaks do not grow back. If the birds have not been beak trimmed properly as chicks, many producers beak trim again before the birds are put into the laying house. Unless the job is done properly, beak trimming at 1 day of age can result in chick mortality due to starve-outs.

Beak trimming can be done at any age to correct a picking problem. However, it is best done before the birds start to lay and should be done by 16 weeks of age.

Beak trimming involves cutting off slightly more than one half of the upper beak and blunting the lower beak. The upper beak should be about ⅛ inch (3.2 mm) shorter than the lower beak, making it difficult for the bird to grasp feathers or skin. A temporary beak-trimming job can be done in the absence of a beak-trimming machine. This is done by removing a small portion of the upper beak with nippers or dikes. The beak soon grows out and the procedure must be repeated.

Properly beak-trimmed pullet

Properly beak-trimmed adult bird

A beak-trimming machine has an electrically heated blade that cuts and cauterizes at the same time to avoid bleeding. Properly beak-trimmed growing and adult birds are shown here.

Lighting Programs for Young Stock

During the first 3 or 4 days, provide the chicks with 22 hours of light. During this time the chicks will get used to their surroundings and learn where the heat, feed, and water are located.

After the initial period, give careful attention to the lighting schedule for chicks grown for layers. Birds are very responsive to light. This response is an excellent example of the interaction of the nervous system with the various endocrine glands. The eye and other, extraretinal photoreceptors, such as the pineal gland and the harderian gland, receive the light stimulus, resulting in the discharge of certain hormone-releasing factors from the hypothalamus, which in turn causes hormonal secretions in other glands. The endocrine glands play a vital role in body processes, including growth, rate of sexual maturity, and rate of egg production.

If growing birds are exposed to an increasing day length, particularly toward the end of the growing period, they will mature sexually at too early an age. The result will be smaller eggs, which are less valuable than the larger sizes. Total egg production may be lower, and there may be a high incidence of prolapse of the uterus and other problems. The rule of thumb to follow is never expose laying birds to a *decreasing* day length, and never expose growing pullets to an *increasing* day length.

At the equator, daylight length remains constant, so lighting schedules are not much of a factor in pullet growth and production. North and south of the equator, daylight length varies seasonally: The farther from the equator, the greater the difference.

Several programs have been used for growing pullets to prevent early sexual maturity and its related problems. Lighting programs are recommended based on the type of housing used. In windowless (environmentally controlled or light-tight) houses, the suggested lighting program usually seeks to maintain a continuous light period after the first few weeks, using a minimum of 1 foot-candle of light at bird level.

One such program, for example, calls for 22 hours of light for week 1, 18 hours for week 2, and 16 hours for week 3. Beginning at 4 weeks and continuing through 16 weeks, the birds are kept on an 8- to 10-hour day.

At 17 weeks through 20 weeks, the light is increased by 1 hour weekly; and from 21 weeks, the light is increased by 30 minutes per week until the day length reaches 15 hours.

Light Requirements for Conventional (Open-sided) Houses

Growing Period

◆ Start chicks on 22 hours of light for the first week of age. Reduce day length to 18 hours in week 2 and 16 hours in week 3. Light intensity during growth should be 5 to 10 lux (0.5–1 foot-candle) at bird level.

◆ From weeks 4 through 16, sex-link commercial pullets should be grown on a constant day length according to the appropriate light requirements for month and latitude in the following table. Light intensity during this period should not exceed 10 lux (1 foot-candle) at bird level.

Production Period

◆ At 16 weeks of age, increase light intensity to 20 lux (2 foot-candles) at bird level. This level should be maintained throughout the production period.

◆ Where constant day length in the growing period was less than 14 hours, increase day length by 1 hour in week 17 and 1 hour in week 18. Continue to increase day length by 30 minutes per week until a maximum day length of 16 hours is reached.

◆ Where constant day length in the growing period was 14 hours or greater, increase light 1 hour in week 17 and add light in 30-minute increments until a 17-hour day length is attained.

◆ Keep clocks on Standard Time and divide day-length increases evenly between morning and evening.

One brown-egg breeder recommends the following lighting program for conventional or open-sided houses.

Pullets grown in open-sided or windowed houses are exposed to natural daylight conditions and are treated somewhat differently. If the birds are hatched during the time of year when daylight length will be decreasing during the growing period (April through July in the Northern Hemisphere), the pullets can be reared under natural lighting conditions. If the chicks are hatched during the August through March period, special lighting programs must be used — either a step-down program or a constant-lighting program.

The step-down program is frequently used in the New England area for August through March hatches. The length of day 24 weeks from the date of hatch is determined. Six hours are added to the length of day at that time and the birds are started on that amount of light. The total light period is then reduced by 15 minutes per week throughout the growing period, and at 24 weeks of age, the birds are on a 14-hour day. Then 30 minutes per week are needed for a few weeks until the total day length reaches 15 to 16 hours. The total hours of light should be slightly more than the number of hours of natural daylight in your area. It should be noted that the light periods in the programs for open-sided or windowed houses are those totals of natural daylight supplemented by artificial light.

Constant Light Program in Growing Period
Conventional (Open-sided) Houses

| HATCH DATE | | HOURS OF LIGHT IN GROWING PERIOD | | |
| | | LATITUDE | | |
NORTH OF EQUATOR	SOUTH OF EQUATOR	0°–29°	30°–39°	40°–45°
January 15	July 15	13½	14½	14½
February 15	August 15	14	15	15½
March 15	September 15	14	15	15½
April 15	October 15	13½	14	14
May 15	November 15	12½	13	13
June 15	December 15	12	11½	11½
July 15	January 15	11½	10½	10
August 15	February 15	11½	10	9
September 15	March 15	11½	10	9½
October 15	April 15	11½	11	10½
November 15	May 15	12	12	12
December 15	June 15	13	13	13½

Example: A flock hatched in Thailand (15° N latitude) on April 15. Maximum natural light during growth will be 13½ hours. Provide 13½ hours of light from 4 weeks of age through 16 weeks of age.

Costs of Growing Pullets

Many factors affect the costs of production. The most flexible will be the cost of feed. Where homegrown grains are available or a range-rearing program is used, substantial savings may be made and the costs will vary from those shown in the chart.

Estimated Cost of Raising Pullet Replacements to 22 Weeks[a]

ITEM	BROWN-EGG PULLET	WHITE-EGG PULLET
Chick cost	$.80–1.20	$.80–1.20
Feed cost @ $.10 per lb[b]	2.00	1.60
Brooding, litter, and misc.	.30	.30
TOTAL	$3.10–3.50	$2.70–3.10

Assumptions: Mortality 5 percent, 20-pound (9.1 kg) feed consumption for brown-egg birds, 16-pound (7.2 kg) consumption for white-egg birds full-fed. Costs vary depending upon quantities purchased, feed program, feed prices, and other factors.

[a] *Interest on average investment, labor, and overhead (depreciation, insurance, taxes) are costs of producing commercial pullets not normally considered in small-flock situations. Estimated costs are interest: 13–15 cents; labor: 25 cents; overhead: 26–28 cents.*

[b] *The feed price shown is averaged between bulk feed and bag feed costs and is provided for illustrative purposes only.*

A HEALTHFUL HOBBY

Having raised chickens as a kid in 4-H, Joe always enjoyed poultry rearing as a hobby. Now, as a family man, he keeps some Barred Plymouth Rocks for eggs, and raises about thirty-five broilers each year to put in the freezer, along with six turkeys — three to keep for the family and three to sell to help cover costs.

He built a coop and an outside pen for his turkeys about 75 feet (22.9 m) from his two chicken coops. One coop has an outside pen and the other is for growing broilers. Each coop has running water and electricity. Joe built a small composter for his litter and household table scraps and uses the compost he generates on his garden. He sells a few eggs to friends and neighbors and keeps the rest for his family. He says this has been a good hobby for him and that it keeps him out of trouble.

SIX

MANAGING THE LAYING FLOCK

The first step in managing your laying flock is the necessary preparation for the birds' arrival. You must be ready for them with clean, well-lighted and well-ventilated housing, with proper feed and water, and with adequate nesting space.

Preparing the Laying House

Thoroughly clean the laying house between flocks — this means removing the manure and litter, and brushing down the cobwebs and dust from walls and ceilings. A good wet cleaning involves scraping all the caked material from the floors and walls, then thoroughly scrubbing and rinsing with a garden hose or steam cleaner. Do any repairs that are needed at this time.

After you have cleaned the house, spray it with a disinfectant. Also clean and disinfect the equipment at this time. Some of the common disinfectants are quaternary ammonium compounds, creosol solutions, hypochlorites, and the phenol or carbolic acid preparations. Use them according to the manufacturer's recommendations.

When you have disinfected the house, paint the roosts, dropping boards, or pits, and paint the nests with carbolineum or a red-mite paint. It is advisable to treat the floors with carbolineum once each year. It is not only a disinfectant but also a wood preservative, and it makes future cleaning much easier. Use a similar cleaning and disinfecting program for the brooder house and equipment.

Caution

Carbolineum is an excellent material but must be used correctly. Allow a minimum of 2 weeks for the house to dry and air out before you put the birds in. When wet, the material can be injurious to their feet and eyes.

Litter

Provide approximately 6 to 8 inches (15.2–20.3 cm) of clean, dry litter material. Although availability and cost are factors in determining the type of litter to be used, some of the more commonly used materials are shavings, sawdust, a combination of shavings and sawdust, ground corncobs, and sugarcane. These are all excellent.

To help maintain the litter in a satisfactory condition, it is recommended that you use a built-up litter. Start it early in the fall, before cold weather sets in. Begin with 4 to 5 inches (10.2–12.7 cm) of clean litter and gradually add to it. Fresh litter is added until a depth of 9 to 12 inches (22.9–30.5 cm) is reached. It is usually not necessary to change the litter during the laying year. Removal of the wet spots from time to time is typically all that is required.

Built-up litter insulates the floor and provides warmth for the birds. Due to the decomposition and fermentation processes, it actually produces heat in the litter. It also absorbs moisture from the feces.

Don't use the same litter for more than 1 year. Infestations of internal and external parasites can result. Then, too, it is necessary to remove the litter to properly clean the house.

When warm weather arrives, it is the practice of some producers to replace the built-up litter with new litter. This provides cooler, more comfortable conditions during hot weather. In cold weather, it is usually necessary to stir the litter regularly to keep it from packing and to permit it to aerate and dry.

Feeding the Layers

Laying birds require a balanced diet containing all the necessary nutrients, such as protein (supplying the essential amino acids), carbohydrates, fat, vitamins, and minerals. Actually, the requirements for these nutrients vary during the laying cycle. For example, protein intake needs to be higher during the early laying period, because that is when egg production peaks and the birds are still growing. As egg production diminishes, the protein requirement decreases.

Protein is expensive, so commercial producers frequently *phase-feed* their layers; that is, they use different protein levels during the laying period. They may start with an 18 percent diet and, at approximately 4 months of production, reduce the level to 16 percent. When the layers fall below 60 to 65 percent production, the protein content of the feed is dropped to 14 or 15 percent. Methionine, one of the essential amino acids, is one of the limiting factors in the diet and is usually supplemented in poultry feeds.

Many factors affect the feed intake of the birds. With changes in feed consumption, protein intake changes, so feeding layers is not a simple task if you are aiming for optimum production. Some of the factors that affect feed intake include:

- Management matters such as feed trough and floor space
- Environmental temperatures, both warm and cold
- Caloric content of the diet
- Variations in egg production
- Flock health or stress

Most small-flock owners prefer to use a simple feeding program, one that uses a single type of feed and can be full-fed to the layers. In these situations, a 16 percent to 17 percent all-mash diet is normally used.

All-Mash Feeding

The simplest and most foolproof approach to feeding layers is to use a complete all-mash laying ration and keep it in front of the birds at all times. It is less bother, is adaptable to mechanical feeding, and provides a more nearly balanced diet for the layers, assuming there is enough feeder space available.

Grain-and-Mash Feeding

Mash is a mixture of finely ground grains. It may be formulated to provide all the nutritive requirements of the birds or to be fed with grain or other supplements. *Grit* refers to hard, insoluble materials fed to birds to provide a grinding material in the gizzard. It may be useful when birds are fed high-fiber feed ingredients, to aid in their digestion. Small stones or granite particles are good grit materials.

A grain-and-mash system of feeding is still preferred by some producers. Grains are frequently fed in the form of so-called scratch feed, containing corn, wheat, and oats. It is often thought to be somewhat easier to maintain a high feed consumption with the grain-and-mash systems. Grains are also well liked by the birds and they eagerly consume them when they are provided.

Some producers feed a large part of the daily allowance of grain before the birds go to roost. During cold weather, this gives them a reserve supply of energy for warmth and will help keep them more comfortable while on the roost during the night. Grains are high in energy and are digested more slowly than mash, thus providing more warmth over a longer period of time. Grit must be provided when hard grains are fed.

Maintaining Protein Intake

The portions of mash to grain vary with the protein content of the mash and also the type of layer. Usually, a 20 to 21 percent mash is kept before the birds at all times. Light breeds such as Leghorns, on a grain-and-mash program, are fed equal amounts of grain and mash, or perhaps slightly less grain than mash. Heavy breeds have a tendency to become too fat if they receive too much energy, and therefore may receive only 40 percent grain and 60 percent mash. In some instances, if a high-energy mash is used, it is necessary to decrease the amount of grain to 30 percent of the total diet. It is important to prevent laying birds from getting too fat. Excessive fat is not conducive to high egg production.

One of the problems encountered with the mash-and-grain system is how to feed in the proportions that will provide the correct protein intake of approximately 16 percent. Production problems can frequently be traced to a low total intake of protein when a mash-and-scratch system is being used.

The protein content of most scratch grains is approximately 9 to 10 percent. If, for example, equal proportions of an 18 percent laying mash and scratch grains are fed, the total protein intake will be approximately 13.5 to 14 percent. This will not support optimum egg production — at least during the early part of the laying cycle. Most high-energy complete laying mashes can safely be supplemented with grain at the rate of 1 or 2 pounds (0.5 or 0.9 kg) per hundred birds per day. Follow the manufacturer's recommendations.

Feeding Tip

You may feed grain on top of the mash or in the litter if litter conditions are reasonably dry and clean. When birds scratch for grains, it helps aerate the litter and maintain its condition.

Free Choice, or Cafeteria-Style

This method uses a high-protein mash containing 26 to 40 percent protein, with grain and calcium supplement kept before the birds at all times. The chief advantage of this method is that it enables you to use local grain or grain produced on your farm. It is also a simpler system than hand-feeding grain. Feed the mash concentrate in one hopper and the grains in individual hoppers.

Other Feeding Programs

If you have a small flock, you can feed your layers fresh kitchen scraps, garden products, and even surplus milk. Feeding these various materials can substantially reduce the amount of purchased grain required. Limit the amounts of these materials to what the birds can clean up in 5 or 10 minutes. If you feed too much of this sort of material to the birds, the nutritive intake can be diluted to the extent that they don't get the right amount of protein and other nutrients for body maintenance and the production of eggs. Give milk to the birds in plastic or enamel, *not* galvanized, containers.

Take care in the choice of table scraps you give the birds. Spoiled meat and materials such as fruit peelings, onions, and other strong-flavored foods may give a bad taste to the eggs and should not be fed to the birds. You can feed the chickens potato peelings if they are cooked. Vegetable peelings and green tops of vegetables are good. Green feeds that are satisfactory for laying birds are similar to those recommended for growing birds. When you supplement a complete laying mash with table scraps and other materials, keep these supplemental feed sources to 10 percent or less of their total daily intake so as not to upset their nutrient balance, and provide the birds with grit and calcium supplements. Eggshells are a good source of calcium and may be crushed and fed back to the laying flock. A simple or "natural" laying hen diet for those who want to mix their own ration can be found on page 299.

Feed Consumption of Layers

Feed consumption varies with the body weight and the rate of production. A certain amount is used for body maintenance. This amount corresponds to the amount consumed at 0 production. As a rule of thumb to estimate feed consumption at various levels of egg production, add to the amount of feed required for maintenance (0 production) 1 pound (0.5 kg) of feed for each 10 percent of egg production.

Birds eat to satisfy their energy requirements, so the feed consumption will vary with environmental temperatures. They will eat more in cold weather. To ensure optimum feed consumption, provide 4 linear inches (10.2 cm) of feeder space per layer.

Calcium Supplements and Grit

Calcium is one of the most important minerals in the layers' diet. Hens need calcium for eggshell formation. About 10 percent of the total weight of the egg is shell, and the shell is almost 100 percent calcium carbonate. Hens in heavy production require calcium in relatively large quantities.

Most of the high-energy complete laying mashes today contain 3 to 3.5 percent calcium, depending upon the time of year. Under most circumstances, this appears adequate to produce eggs with sound shells.

Feed Consumption at Various Levels of Production*

PERCENT PRODUCTION	LEGHORN-TYPE BIRD — 4 LB		BROWN-EGG BIRD — 5 LB		BROWN-EGG BIRD — 6 LB	
	LB/DAY	LB/DOZ	LB/DAY	LB/DOZ	LB/DAY	LB/DOZ
0	15.8		18.6		21.2	
10	16.7	20.1	19.6	23.6	22.1	26.6
20	17.6	10.5	20.4	12.2	22.9	13.7
30	18.4	7.4	21.3	8.5	23.9	9.6
40	19.3	5.8	22.2	6.7	24.7	7.4
50	20.2	4.8	23.1	5.5	25.6	6.1
60	21.1	4.2	24.0	4.8	26.5	5.3
70	22.0	3.8	24.9	4.3	27.5	4.7
80	22.9	3.4	25.8	3.9	28.1	4.2
90	23.8	3.2	26.7	3.6	29.2	3.9

*Feed per day per 100 layers and feed per dozen eggs.

Data from Dustic, R. E., and M. C. Nesheim, Poultry Production, 13th ed. Philadelphia: Lea & Febiger, 1990.

When laying birds get well into their laying cycle and shell quality begins to deteriorate, or during periods of extremely hot weather when shell quality may suffer, additional calcium seems to be needed. In these cases, producers commonly add extra calcium to the diet in the form of oyster shells. Some types of birds, particularly the light-laying strains that have relatively small calcium reserves, need calcium supplementation earlier in the laying cycle than some of the heavier types of birds. When to feed supplemental calcium and when not to is still somewhat controversial. Some feel that the most satisfactory way of meeting the variable individual requirements of layers is to supplement the calcium in the laying mash. You can do this by feeding separately hen-sized oyster shell or calcite crystals, both of which are soluble.

For feeding systems other than the complete laying mash system, a supplemental feeding of calcium is recommended.

In addition to a source of calcium, a hard, insoluble grit should be fed in many instances. Insoluble grit, such as granite grit, maintains its sharp edges in the digestive system and in essence takes the place of teeth. It is often thought that it is not necessary to feed grit to layers fed an all-mash diet. It is assumed that everything the birds consume in the laying diet is already finely ground and, therefore, needs little or no further grinding. Where birds are kept in cages, this is probably true.

Birds housed in a floor management system frequently eat feathers, litter, and nesting material. Under these circumstances, grit, if available, enables the gizzard to grind the fibrous material so that it can be moved on through the digestive system. Any feeding program other than the all-mash system should incorporate grit so coarse grains can be utilized by the digestive system.

Hoppers

There are a number of types of grit and shell hoppers available on the market. These are typically divided into two compartments so you can feed both grit and the source of calcium from the same hopper. It is a relatively simple project to construct a hopper that will serve equally well. In those situations where grit and calcium material are needed, heap the hopper full at all times.

Watering

Birds need plenty of clean, cool water at all times if they are to produce well. Water makes up a large portion of the hen's body and is a major constituent of the egg. Water helps soften the feed and aids in its digestion, absorption, and assimilation. If hens are deprived of water for only a short time, egg production will suffer. Dirty water and watering equipment may discourage water consumption and are potential sources of disease infection. Never allow your water supply to freeze for an extended period. If you are unable to maintain the house temperature at a level sufficient to prevent freezing, you can install an electrical immersion heating unit in the drinking fountain or use pan heaters. Heating units should be thermostatically controlled.

Laying birds normally drink in pounds about twice as much as the feed they consume. During hot weather, they will consume substantially more than this amount. Provide 1 inch (2.5 cm) of the trough type of waterer per bird, or one round waterer per one hundred birds. One pan type of waterer is usually adequate for the small family flock. Clean and fill the waterers with fresh water daily. The water consumption of layers is shown in the chart.

Water Consumption of Layers Based on Environmental Temperature

Temperature	Gallons per 100 Layers per Day
20–40°F	4.2–5.0
41–60°F	5.0–5.8
61–80°F	5.8–7.0
81–100°F	7.0–11.6

Source: New England Poultry Management and Business Analysis Manual, *Bulletin 566 (Revised).*

Facilities and Management

In well-insulated, mechanically ventilated houses, you can house Leghorn layers at the rate of 1¼ to 1½ square feet (0.12–0.14 sq m) per bird. Heavier brown-egg birds should be given 1½ to 2 square feet (0.14–0.19 sq m) per bird. In those houses where gravity ventilation is used — that is, ventilation through windows or slots — allow slightly more floor space. The normal recommendation is a minimum of 1½ to 2 square feet (0.14–0.19 sq m) for Leghorns and 2 to 2½ square feet (0.19–0.2 sq m) for brown-egg birds.

Actually, in a well-insulated house, cold-weather ventilation is somewhat simplified if there are more birds, since they supply more heat. If you locate the feeders and waterers over a dropping pit, you can reduce the floor area per bird below the amounts previously noted. However, with each additional reduction in floor space per bird, you must provide additional feeding, watering, nesting, and roosting space. The birds should not have to travel more than 15 feet (4.6 cm) to reach either feed or water.

Nests

Provide at least one individual nest or 1 square foot (0.09 sq m) of community nest space for each four hens. Replenish nesting material as needed and keep it clean and dry. This is important in order to prevent dirty, cracked, and broken eggs. The nest should be a minimum of 2 feet (0.6 m) from the floor or litter. If floor eggs are a problem at the onset of production, it is advisable to place the nests on the floor and gradually raise them as the birds become accustomed to using them. One of the common problems with small flocks is the habit of egg eating. If the nests are poorly constructed or without litter, or the eggs are not gathered frequently enough, breakage occurs, and egg eating may result.

Egg Gathering

Gather the eggs two or three times a day. Frequent gathering improves the egg quality and reduces the number of dirty eggs as well as the likelihood of birds' developing the habit of egg eating. Often it is difficult for small-flock owners to gather the eggs as frequently as is desirable. Having plenty of nests, and keeping them clean and with generous amounts of nesting material, will help prevent broken or dirty eggs. When you have to be away during the day and are unable to collect the eggs, it is possible to set the lights to come on earlier in the morning, so that some of the eggs will be laid earlier and can be gathered before you leave for the day. Eggs not gathered during the day are also subject to freezing in cold weather.

Helpful Hint

Once egg eating commences, it is a difficult habit to stop. Probably one of the best methods of preventing it is to beak trim the birds and increase the frequency of egg collection.

Lighting

The use of artificial light in the poultry house is not intended to give the hens more time to eat. It is the stimulation of the light itself that makes them lay more eggs. The light stimulates the pituitary gland through the eye. This gland, in turn, secretes hormones that stimulate the ovary of the hen to lay eggs.

Provide laying birds with 15 to 16 hours of light daily during their laying period. In the northern part of the United States, darkness exceeds the amount of daylight during many of the fall and winter months. Beginning on August 15, birds in windowless houses need supplemental light to give them a 15- to 16-hour day. During the months of November, December, and January, when the days are shortest, they will need as much as 7 hours of supplemental light. For best results, keep the day length static by providing supplemental lights prior to sunup and turning them off after sunup and on again prior to sundown and leaving them on until a total of 15 or 16 hours of light has been achieved. A multi-setting time clock is ideal for this chore.

Layers housed in controlled-environment or light-tight houses will need a constant 15 or 16 hours of light.

Lamp Types and Placement

For supplemental lighting, lamps can be either 40- to 60-watt incandescent or 7- to 13-watt compact fluorescent. Use a reflector to get the most efficient use of the lamps. Install one light fixture for each 200 square feet (18.6 sq m) of floor area or less. A distance of 10 feet (3.1 m) between lamps generally provides a good distribution of light for the birds. A rough rule of thumb for light intensity is 1 foot-candle at the feeder level. For incandescent lamps, 1 bulb watt per 4 square feet (0.4 m) of floor space will usually provide about 1 foot-candle if the bulb is 7 to 8 feet (2.1–2.4 m) from the litter and has a reflector. A light dimmer can be used to control light intensity, to provide bright light for collecting eggs and cleaning and dim light for the birds. During the first week or two of the lighting period, it may be necessary to catch and place some of the birds on the perches if they do not get onto the roost of their own accord when it's time for lights-out. It is well worth the time and effort you spend teaching the birds to use the roosts.

Ventilation

During cold weather, all of the ventilation should be from one side of the poultry house, preferably the south. In the absence of fan ventilation, windows that slide down from the top or tip in from the top are best for winter ventilation. Adjust the windows according to the weather. When the house tends to be stuffy and the ammonia fumes are strong, it needs more ventilation. The house should never be closed tight, even on cold nights. Always leave at least some of the windows slightly cracked.

In the warm summer months or in warm climates, ventilation will help cool the house. If you can open both of its sides, it will increase the air movement, help keep the birds cool, and maintain better egg production and overall performance.

Rodent and Bird Control

Rats and mice like chicken feed. If there is a place for them to hide in the chicken house, they will do it. They enjoy living in piles of refuse, the double walls of buildings, old stone foundations, or dropping pits, under deep litter or under the floors, and in many other hiding places. Good housekeeping will help reduce the problem of rats and mice.

Anticoagulant and other types of rat poisons have been very helpful in getting rid of rat problems. For an anticoagulant poison to be effective, rodents must consume it for several days. If you house birds in an old building that has plenty of hiding places, it may be best to set up a permanent bait station or stations, and keep bait in them at all times. See the chapter on flock health for more information on rodent control.

Screen wild birds out of the chicken house, if possible. They are vectors of some of the common diseases and parasites of poultry.

Culling and Selection

Good production-bred chickens will lay well for approximately a year if they are well managed. Culling healthy birds that go out of production for just a short time is not advisable, unless a chicken dinner is in order. After the birds have laid for nearly a year, the nonlayers may be culled, dressed, and used at home or sold. Usually, you should replace the entire flock

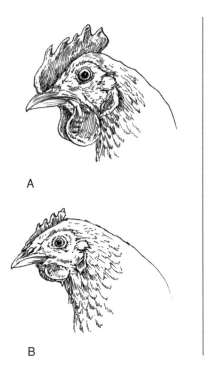

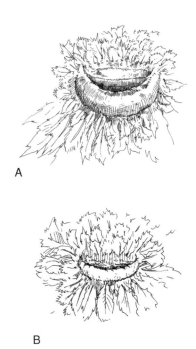

A. *Comb/wattle of bird in laying condition.* B. *Comb/wattle of bird* not *in laying condition.*

A. *Vent of bird in laying condition.* B. *Vent of bird* not *in laying condition.*

after 12 to 14 months of production. On occasion, it is necessary — and advisable — to sell hens that are still laying at a reasonable rate, to make room for the new flock of pullets.

In culling, there are two distinct things that you must learn. One is how to distinguish between a layer and a nonlayer. The other is how to determine with a reasonable degree of accuracy how long a bird has been either in or out of production. After you have acquired the ability to do these two things, you can do a reasonably good job of culling a flock of layers.

With the beginning and termination of laying, changes take place in the head, the abdomen, and the vent. To determine whether or not a layer is in production, all three of the body parts should be carefully examined to make an accurate judgment. The examination of one of these parts in isolation will not provide the answer.

When birds are in laying condition, their combs and wattles are enlarged, bright red, and waxy in appearance. The vent is large and moist, and the pubic or pin bones become flexible and are wide apart to physically permit the egg to be laid. The spread of the pubic bones of the laying bird will be about three fingers wide. The abdomen will be soft and pliable.

Characteristics of High and Low Producers

CHARACTER	HIGH PRODUCERS	LOW PRODUCERS
Vent	Bleached, large, oval, moist	Yellow, dry, round, puckered
Eye ring	Bleached	Yellow-tinted
Beak	Bleached or bleaching	Yellow or growing yellow
Shanks	Pale yellow to white, thin, flat	Yellowish, round, full
Head	Clean-cut, bright red, balanced	Coarse or overrefined, dull, long, flat
Eyes	Prominent, bright, sparkling	Sunken, listless
Face	Clean-cut, lean, free from yellow color and feathers	Sunken or beefy, full, yellowish, feathered
Body	Deep	Shallow
Back	Wide; width carried out to pubic bones	Narrow, tapering, pinched
Plumage	Worn, dry, soiled	Smooth, glossy, unsoiled
Molt	Late molter	Early molter
Carriage	Active and alert	Lazy and listless

Characteristics of Layers and Nonlayers

CHARACTER	LAYING HEN	NONLAYING HEN
Comb	Large, red, waxy, full	Small, pale, scaly, shrunken
Wattles	Large, prominent	Small, contracted
Vent	Large, moist	Dry, puckered
Abdomen	Full, soft, velvety, pliable	Shallow or full of hard fat
Pubic bones	Flexible, wide open	Stiff, close together

Source: Culling for High Egg Production, *Vermont Agricultural Extension Service Circular 115RU.*

When birds are not laying, the combs and wattles may be small, pale, and scaly in appearance, the vent dry and puckered. The pubic bones may be close together (1 or 2 fingers in width), and the abdomen hard and shallow.

The characteristics of high and low producers and layers and nonlayers are presented in the charts on page 110.

To determine how long a bird has been in production or out of production, you must consider two factors; namely, the degree of bleaching or pigment and the molt.

Pigmentation

The yellow pigmentation found in all yellow-skinned breeds and varieties of chickens is a pigment called *xanthophyll*. Yellow corn, a major constituent of the poultry diet, is the principal source of this pigment. If a pullet has been properly fed when it comes into the laying house, it will carry a considerable amount of pigmentation in all parts of its body. When it commences to lay, the pigment present in the skin and in other parts of the body is gradually lost. The reason for this loss is that the pigment is diverted to the yolks, giving them their yellow color. As long as the bird is in production, it continues to lose this yellow color from the various parts of its body. Hens that are thoroughly bleached out are usually high producers. When the bird stops production, the pigment reappears in the various parts of the body. The return of pigmentation is one of the first clues that a bird is out of production.

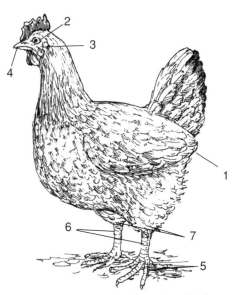

Pigment is bleached from the skin in this sequence; (1) vent, (2) eye ring, (3) earlobe, (4) beak, (5) bottoms of feet, (6) shanks, and (7) hocks.

There is a definite order of bleaching of the body parts as the birds begin to lay — first the vent, then eye ring, earlobe, beak, feet, shanks, and hocks. The order of bleaching and the approximate time required to bleach the various parts of the body for birds in continuous production are given in the chart.

Bleaching Sequence and Time Required

BODY PART	TIME REQUIRED
Vent	Few days
Eye ring	2 to 3 weeks
Earlobe	3 to 4 weeks
Beak	6 weeks
Bottoms of feet	10 to 12 weeks
Shanks	2 to 5 months
Hocks	6+ months

Vent

The first noticeable change occurs in the skin around the edges of the vent. A few days after production begins, the yellow color around the vent disappears. A yellow vent indicates that a bird is *not* laying. A whitish, pinkish, or bluish white vent indicates that a hen *is* laying.

Eye Ring

The eye ring starts to bleach soon after the vent. In 2 to 3 weeks after the onset of production, it usually loses its yellow pigment. By that time, the bird has probably laid ten to twelve eggs.

Beak

The yellow color leaves the beak next. It leaves the base of the beak first, and the fading continues toward the tip. It takes approximately 6 weeks of continuous production for complete bleaching. By that time, usually thirty to forty eggs have been laid.

The lower beak bleaches more rapidly than the upper beak. Fading of color can be seen readily in the lower beak. When the dark pigment "horn" appears on the upper beak, the lower beak can be used as a basis for judging the degree of pigmentation loss. Barred Plymouth Rocks and Rhode Island Reds commonly carry this dark pigment.

Shanks

The bleaching of the shanks occurs last and is an indicator of long-term production. Bleaching of the shanks takes from 4 to 5 months of continuous production. The bird must lay from 120 to 140 eggs to be completely bleached. It is not possible to pick a bird as a layer or a non-layer on the basis of bleached or yellow legs unless it has been in production for 4 or 5 months.

When the bird stops laying, pigmentation returns to the parts of the body in the same order in which it bleached out.

By observing the degree of pigmentation, you can tell rather accurately how well a bird is laying. You must take into consideration the fact that the feeding program, breed, and flock health are factors that can also affect pigmentation.

The Molt

Once each year birds renew their plumage. This process of replacing old feathers with new ones is called *molt*. Hens usually go through their annual molt in the late summer, the fall, or early-winter months. Factors that determine the time of the molt are:

1. Time of the year the bird was hatched.
2. The individual bird or breeding.
3. Management stresses to which the bird is exposed.

Just as bleaching is most helpful in determining layers from nonlayers during the first 8 or 9 months of production, the molt is most useful during the last several months.

When a bird starts its molt, it goes out of production and its reproductive tract shrinks to the size of that of an immature pullet. The bird will not come back into production until shortly before or just after the molt is completed and the reproductive tract returns to normal size. The pattern depends to a certain extent upon the type of management and feeding program to which the bird is exposed. Some laying birds start their molt early in the fall after 8 or 9 months of production and are called *early molters*. Others lay for 12 or 15 months before they molt and stop production. In fact, some of our modern-day strains of layers have to be force-molted to get them out of production. This procedure provides them with an opportunity to rest, renew their feathers, and return to production. Hens that complete at least 12 months of production before molting are referred to as *late molters* and are the most desirable birds to have in the flock.

Primary Feathers

When using the molt to cull or select hens, use the primary (flight) feathers of the wing. There are usually ten of these feathers, arranged in a group, extending from the short axial feather in the center of the wing to the tip.

The fact that the primary feathers are dropped and renewed in a definite order makes the use of the molt a fairly accurate means of estimating how long a bird has been out of production and how long it will take to go through its complete molt. It is also possible to determine whether the bird is a slow molter or a rapid molter by observing the way in which it has dropped and is renewing the primary feathers. A typical slow-molting bird will drop one primary at a time over an extended period, whereas a rapid-molting hen will drop more than one at a time and drop them more frequently. In some of the better strains of laying birds, the hens will drop a group of several primaries and then, very soon after, drop more, and thus go through the complete molt in a relatively short period of time and be back in production. Approximately 6 weeks is required to grow a primary feather.

Usually, the slow molters stop production before or at the time they start their wing molt. They seldom lay through a molt. Thus, the slow molter that drops one primary at a time will be out of production for many weeks and would be a likely candidate to be culled from the flock. Many of the rapid-molting birds will lay for a period of time after they start to renew their primary feathers. Some of the good laying strains of birds will renew half or more of their primaries before they finally stop laying. The illustration shows the different stages of molt in fast- and slow-molting birds.

Wings during Different Stages of Molt

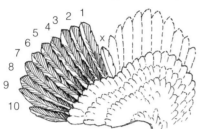

The ten old primary feathers *(black)* and the secondary feathers *(broken outline)* are separated by the axial feather *(x)*.

A slow molter at 6 weeks of molt, with one fully grown primary and feathers 2, 3, and 4 developing at 2-week intervals.

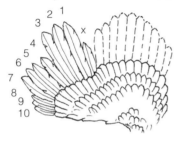

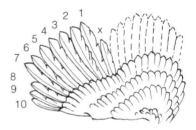

In contrast, a fast molter has all new feathers. Feathers 1 to 3 were dropped first (now fully developed); feathers 4 to 7 were dropped next (now 4 weeks old); and feathers 8 to 10 were dropped last (now 2 weeks old).

Two weeks later, feathers 1 to 7 are fully grown. Fast molt took 10 weeks, compared with 24 weeks for slow molt.

(Redrawn from South Dakota Extension Service.)

Force Molting

Toward the end of the laying cycle, egg production is at a relatively low level. Interior and exterior egg quality is poorer, particularly if the tail end of the cycle coincides with hot weather. At this point, many producers force-molt or recycle the flock if it has been a good one.

Force molting gives the layers a rest for about 8 weeks. After the rest, egg production increases, but not to the level of the pullet year. It may reach 88 to 90 percent of the pullet-year level. Egg quality, both exterior and interior, recovers substantially.

Recycled layers usually lay profitably for only 6 to 9 months. Egg size is large, feed consumption is higher, and the overall egg quality is lower than during the pullet year. In some cases, mortality will be slightly higher than in the pullet year.

The main reason for force molting is to beat the high cost of pullet depreciation, which may be 8 to 10 cents per dozen, depending upon the pullet cost and the fowl or salvage value. If force-molted birds are to be profitable to the commercial producer, there must be a market for the extra-large eggs. The profitability of the practice and your decision to molt depends on many factors, including pullet cost, price of eggs, and price of feeds.

Force-Molt Procedure

Day 1   Weigh a sample of the birds and obtain an average. Decrease lights to 8 hours per day in light-tight houses or to no artificial light in window houses. Remove feed, but provide water free-choice.

Day 8   When body weight has been reduced by about 30%, provide growing mash at 40–50% normal consumption level.

Day 22   Restore lights to premolt level and full-feed laying ration.

Modified from New England Poultry Management and Business Analysis Manual, *Bulletin 566 (Revised).*

Broodiness

Poultry geneticists have done a great deal to eliminate the natural tendency for laying birds to become broody. There will, however, still be an occasional bird in floor management systems that will become broody. If you let the bird stay in the nest, it will remain broody and will not eat or drink properly during this time, and a loss in body weight and egg production will result. If you remove broody hens from the nest as soon as they are discovered, you can substantially reduce the loss of egg production.

The best procedure for breaking up broody birds is to place them in a broody coop — constructed with a wire or slatted bottom to discourage setting and inactivity. The broody coops are usually suspended from the ceiling in the pen or outside it, and the birds are provided with plenty of feed and water. Birds exposed to this type of treatment can usually be put back with the flock in 4 or 5 days.

Flock Health

Flocks that are well managed and fed a well-balanced diet do not experience many disease problems. If one bird dies now and then, it is probably nothing to become alarmed about. If a number of birds become sick, respiratory symptoms appear, or they look droopy, go off feed, or stop laying, it is then time to find out why.

If a disease outbreak occurs, the cause should be determined as soon as possible. First signs are often a change in feed and water consumption. If you do not recognize the disease or parasite, take or send some of the live birds showing symptoms to the nearest poultry diagnostic laboratory. Poultry diagnosticians prefer live birds showing symptoms. Dead birds sent to the laboratory should be kept cool, but not frozen, so that they do not arrive at the laboratory in a decomposed condition. It is helpful to poultry diagnosticians to know the flock history. Relate the symptoms you have observed in the flock, the number of birds affected, the number that have died, source and size of the flock, feeding program, age of the flock, and any other information that you think will be helpful.

Disease Control

To control disease on the farm, incinerate dead birds or put them into a disposal pit or compost them, and remove all obviously sick chickens from the flock. It is sometimes best to kill them and dispose of the carcasses. If, however, the birds are to be treated, put them in a separate pen or hospital pen — as far away from the other birds as possible.

The old adage "An ounce of prevention is worth a pound of cure" certainly applies to poultry-flock health. If you are successful in buying or raising good healthy pullets, they are properly vaccinated, placed in cleaned and disinfected quarters, given enough feed, water, and floor space, and are fed a well-balanced diet, you are well over the hurdle with regard to disease or production problems. (See chapter 13 for more on flock health.)

Vaccination

Birds should be vaccinated for such diseases as infectious bronchitis, Newcastle disease, fowl cholera, laryngotracheitis, and fowl pox, if these are a problem in your area. A good laying-flock manager checks for internal and external parasites and immediately commences control measures as indicated. Screening windows, doors, and other points of entry for wild birds will help prevent some disease and parasite problems.

Bad Habits

Cannibalism, feather picking, and egg eating are habits that are sometimes hard to stop once they get started. Some of the problems with cannibalism and feather picking occur in the spring, prior to mating, and may be due to increased day length, extra light, and extra time for getting into trouble. Many young birds going through puberty also increase this activity. These behaviors are more common in some breeds of chickens than in others. Different types of feed may also cause these problems. Crumbles and pellets can bring on the picking problem when fed instead of mash. The birds fulfill their intake needs more quickly on pellets and have more time to get into mischief. Proper beak trimming is one answer when those problems begin. Actually, the birds should be properly beak trimmed by 16 weeks of age, so that it doesn't have to be done after the birds come into production. Beak trimming after production commences can cause a production slump.

Frozen Combs

Frozen combs are the small-flock hazard in cold climates. Some breeds and varieties, such as the Single Comb White Leghorn, and the

---

Helpful Hint

Chickens are creatures of habit; any sudden changes can cause them stress. This is true of layers in peak production and those nearing the end of their laying cycle. Any change in feed or management should be made gradually to avoid stress problems.

---

males of many breeds have large combs. These are quite easily frozen in cold housing and can cause lowered egg production and a reduction of egg fertility in mated flocks. When severely frozen, the top of the comb becomes discolored and eventually may slough off. Naturally, there are sore heads, and feed consumption may be discouraged by using grills or reels on feeding and watering equipment (see chapter 2).

You can prevent this problem by *dubbing,* or removing the combs from, the birds. The best time to do this is at the hatchery or the farm at 1 day of age. It is done with a pair of manicure scissors. Cut off the comb close to the head, cutting from the front to the back, with the curve of the scissors up, giving a closely cropped comb at maturity, one that is less likely to be frozen or injured. Clean equipment in a mild chlorine solution between birds, 10 drops of chlorine bleach per quart of water. Dubbing can be done to older birds, at which time both the comb and the wattles are cut with dull shears. Considerable bleeding occurs in older birds. It is not recommended that birds grown in warm or hot climates be dubbed, as the comb is part of the bird's thermoregulatory system.

Records

Record keeping is not only interesting but also necessary if you want to know what the flock is doing and the costs and returns involved. Records need not be complicated but merely contain information on daily egg production, mortality, culling, feed consumption, and the quantity and value of poultry and eggs eaten and sold.

Records may be kept on a calendar or on a pen record obtainable from most feed companies. You can make your own pen record. At the end of the month, the information can be transferred from the pen record to a permanent record. These records should be analyzed each month and at the end of the year. Records should be filled in each day to be accurate. Without good records, it is hard to tell if you should continue in business or buy your eggs and poultry at the store.

Costs of Egg Production

Budgets for small flocks can vary substantially from the information in the following charts. Overhead costs, interest, and labor are frequently not considered in the small-flock situation. Bird depreciation may vary considerably. The fowl may be processed and used at home or sold, and would thus be valued at the consumer price instead of the farm price for

Cost-per-Dozen Analysis

ITEM	COST IN CENTS
White Eggs	
Feed (3.71 pounds) at $10 per cwt*	37.10
Bird depreciation	12.40
Overhead	2.82
Other	1.12
Interest	2.11
Labor	3.72
TOTAL	59.27
Brown Eggs	
Feed (4.28 pounds) at $10 per cwt*	42.80
Bird depreciation	11.80
Overhead	3.76
Other	1.45
Interest	2.66
Labor	3.72
TOTAL	66.19

Feed price shown is averaged between bulk feed and bag feed costs and is provided for illustrative purposes only.

Data from the New England Poultry Management and Business Analysis Manual, *Bulletin 566 (Revised).*

fowl, which is relatively low. Feed price is another variable. Small-flock owners normally buy their feed in small quantities at a price well above that paid by the commercial producer. Feed prices also will depend on the feeding program, the geographical area, ingredient supplies, costs, and other factors. Generally speaking, bag feed prices are typically 30 to 50 percent higher than bulk prices.

The two budgets are based on the following assumptions:

- Birds are housed at 22 weeks of age, and sold at 73 weeks of age (51 weeks of production).
- Annual mortality is 7 percent.
- Eggs per henhouse are 253.
- Feed per 100 birds daily for brown-egg birds is 25.2 pounds (11.4 kg; 90.5 pounds [41 kg] annually).
- Feed per 100 birds daily for white-egg birds is 21.9 pounds (9.9 kg); 78.5 pounds [35.6 kg] annually).

SIMPLICITY, SUNSHINE, AND FRESH AIR

Bob W. has been raising and showing chickens for nearly 60 years and has loved every minute of it. He maintains about 150 birds of a dozen breeds of standards and bantams as well as some interesting crosses he himself has made. Bob lives on approximately 7 acres of land and has many small coops with individual outside runs for his birds. His coops range from modified doghouses to full-size walk-in coops. He believes in simplicity, sunshine, and fresh air in managing his birds, and is successful in rearing healthy ones.

Bob loves to experiment with various crosses and color patterns, and over the years he has developed some interesting-looking chickens, none of which you will ever see in the *Standard of Perfection*. Although Bob has been retired for many years, his birds keep him active and feeling young. He enjoys attending the 4-H youth poultry shows and talking to the kids about poultry.

SEVEN

THE PRODUCTION OF FERTILE EGGS

A rooster is not needed in the laying flock — the hens will lay just as well without it. Some individuals do, however, like to mate their flock of layers merely to produce fertile eggs. Others plan to perpetuate a breed or strain of bird as a hobby, for exhibition purposes, to sell to other poultry fanciers, or for a number of other reasons.

If flocks are mated to produce eggs for hatching, it is good to check on state regulations. Some states require that all hatching-egg flocks be blood-tested for pullorum disease and fowl typhoid. Blood-testing is supervised by employees of the official state agency that administers the National Poultry Improvement Plan. For further information on blood-test requirements, check with your state department of agriculture, state Extension Poultry Specialist, or county Extension agent.

As is the case with other laying flocks, careful feeding and management, good housekeeping, and proper equipment are important if one is to get a good hatch of healthy chicks from the eggs.

Select a source of stock that is healthy and bred to meet the specifications you have set, whether it is feather color, body size, egg production, meat production, or a combination of these characteristics.

Mating

A hatching-egg flock of Leghorn or a Leghorn type of bird should contain about eight or nine males per one hundred pullets, or a ratio of about one male per eleven to thirteen pullets. Heavier breeds, such as Rhode Island Reds and Plymouth Rocks, are mated at the rate of eight males per one hundred pullets. For heavier meat breeds raised on litter, place about ten males per one hundred; and if raised on slats, use a ratio of eleven per one hundred. The ratio of males to pullets in some of the commercial hatching-egg flocks may vary from this ratio substantially, due to anticipated mortality, morbidity, or other special problems with the males. It is always a good idea to add a couple of extra males initially to allow for early culling and losses due to fighting.

Keep a close eye on the males during the breeding season. Males establish a harem and mate with certain females; and if a particular male is unable to mate for some reason, its females will not normally mate with other males until it is removed from the flock.

When handling males, take care not to injure their legs and feet. Always catch them carefully by both legs, or by both legs and a wing.

If you are trying to perpetuate a breed, select the breeding stock to conform to those characteristics desirable for that breed or variety, as outlined in the *American Standard of Perfection,* issued by the American Poultry Association. It describes the various classes, breeds, and varieties of poultry recognized as standard-bred. Primary breeders select for numerous economic characteristics, including production, production efficiency, egg size, egg quality, and livability.

If yearling hens are to be mated, force them out of production to permit approximately an 8-week rest before they commence laying hatching eggs. The same force-molting procedure is used as outlined in chapter 6.

The flock should be mated at least 2 weeks before hatching eggs are to be saved.

A rooster with his harem

Feeding and Management of Breeders

In addition to producing eggs efficiently, breeding stock should produce eggs that hatch well and produce vigorous chicks. Breeder flocks should be fed a diet that has additional vitamins and minerals for high fertility and hatchability. For best results, then, feed the breeding flock a specialized diet formulated for hatching-egg production. These diets are known as *breeder rations*.

The management of a breeding flock is similar to that of a market-egg flock. Most of the recommendations given in the preceding chapter on laying-flock management, therefore, need not be repeated in this chapter. Only the basic differences between breeder-flock and laying-flock management will be mentioned.

One of these basic differences is the management system required. Chapter 2 describes several management systems available for laying birds — among them, the cage system, the all-slat or all-wire, the combination slat-and-litter or wire-and-litter, and the conventional litter management approach.

Cage management systems for breeding birds are relatively uncommon. This is because of the difficulty in designing cages that allow birds to mate with good results. Where cage systems are used, it is usually necessary to artificially inseminate the females. This is not a difficult task where small flocks are involved, but it is much more time-consuming than the conventional floor-pen matings.

Other management systems have been used with some degree of success, but there are a number of problems involved with systems other than the litter floors. This is particularly true with the heavier birds that appear to be inherently lazy and don't want to lay well in the nest. For the small flock, the expense of setting up a two-thirds slat, one-third litter floor facility, as is mostly used by commercial breeders, can be prohibitive. It is probably best not to use wire floors for mated flocks, particularly the heavy-breed flocks, because of the incidence of foot and leg problems and the relative difficulty with which birds mate.

Lighting the Breeders

Light is important, not only to stimulate egg production in breeding flocks but also to increase the semen output of the males. Lighting schedules are outlined in the preceding chapter. Schedules vary with the type of house. Give careful attention to lighting.

Hatchability and Shell Quality

Hatchability of fertile eggs is affected by many factors. Shell quality is one of them. The longer the bird is in production, the poorer the shell quality. Shell quality is affected not only by the length of time the birds have been laying, but also by the time of year, temperature, diet, and genetics. Birds that have been force-molted and are in a second year of egg production invariably show shell-quality deterioration much more rapidly than during the first laying cycle.

Tremulous or loose air cells also affect hatchability; use care in handling eggs to prevent this. Cracked eggs seldom hatch, so take care to produce as few cracks as possible and make sure that cracked eggs are not set in the incubator.

Care of Hatching Eggs

Hatching eggs should be gathered three or four times a day. When temperatures are extremely high or low, they should be gathered more frequently. Eggs should not be left in the nest overnight, and they should be cleaned as soon after gathering as possible.

Cleaning

Eggs for hatching should be clean from the nest. Dirty eggs picked up from the floor or the nests should not be used for incubation. Eggs with a slight amount of dirt should be dry-cleaned by wiping with a dry towel, *not* washed. *Never* use a wet towel, because it can spread disease-causing organisms, and *never* sand hatching eggs. Sanding can make a fine dust containing microorganisms that can settle on other eggs; and sanding also removes the protective cuticle layer, opening shell pores to the environment.

If you have no choice but to use floor eggs or dirty eggs, wash them at a temperature of 110 to 115°F (43–46°C), in water containing an approved egg detergent-sanitizer. Don't immerse them in the wash water for more than 3 minutes. Air-dry the eggs and move them to cold storage.

Improperly cleaned eggs frequently become contaminated and may explode during incubation.

Clean eggs can be sanitized by spraying them with a quaternary ammonium sanitizer mixed in lukewarm water at 200 parts per million (1.3 ounces [36.8 mL] per gallon [3.8 L] of water). A better mixture for spraying eggs as they lie on flats is 1 ounce (28.3 mL) of formalin (40 percent), 1 ounce of quaternary ammonium compound, and 1 gallon (3.8 L) of lukewarm water. Wear rubber gloves and do not breathe the fumes.

Storage

Store fertile hatching eggs at a temperature of 51 to 63°F (10–17°C). Lower storage temperatures may reduce hatchability. Eggs held for less than 5 days should be maintained at the 60 to 63°F (15–17°C) range, while eggs being held longer than 5 days should be maintained at 51°F (10°C) for optimum hatchability. The temperature of the average household refrigerator runs below that recommended for storing hatching eggs, so adjust it accordingly. For best results, keep the relative humidity at 75 percent.

Eggs that are cracked, have thin shells or shells with ridges, or that are excessively dirty or abnormal in size or shape should not be kept for hatching. Excessively large or small eggs are often infertile or won't hatch, and should not be set in the incubator.

Optimum hatchability has been found after storage for 2 to 3 days prior to incubation. Although you may not have enough eggs after only 3 days, for best results hatching eggs should not be stored for more than 10 days to 2 weeks before they are set. They should be stored in filler flats or egg cartons with the small end of the egg faced down. Eggs that are to be held for several days prior to hatching should be held in a slanted position (approximately 35 degrees) and turned at least twice each day. One simple method of turning eggs

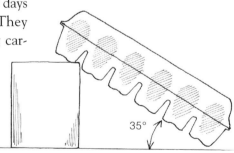

35°

Store hatching eggs at a 35-degree angle for best hatchability.

is to prop up one end of the egg carton or case on an object at an angle of about 35 degrees. Then just shift ends with the container at least twice each day.

Incubators and Operating Temperatures

There are two types of incubators in use today. These are the forced-draft and the still-air machines. The one more frequently used in commercial hatcheries is the forced-draft incubator. This type has fans that force the air through the machine and around the eggs. For most types of eggs, the incubator temperature of the forced-draft machine is set at 99.5 to 99.75°F (37.5–37.6°C). For days 20 and 21, the temperature is reduced to 98 to 99°F (36.7–37.2°C).

Most of the still-air incubators in use today are quite small. They are made to hold from one to one hundred or more eggs. Still-air machines do not have fans. They depend upon gravity ventilation through vents on the top and bottom. The operating temperature of the still-air machine is higher than for the forced-draft incubator. It ranges from 100.5 to 102.75°F (38–39°C), depending on the type of egg being incubated. Actually, you may obtain good results by using a constant 100.5°F (38°C). Calibrating the incubator prior to use is important. Thermostats on most inexpensive, still-air machines will maintain temperature within a range of plus or minus 1 to 1.5°F. You want the average temperature to be what the manufacturer suggests. Thus, determine the highest and lowest temperatures, add them, and divide by two to get the average. Adjust as necessary. For days 20 and 21 of incubation for chicken eggs, drop the temperature by about 1°F.

Ventilation

Ventilation is important in the incubator. Without it the embryos will suffocate. For machines with adjustable vent openings, the vents are usually cracked open at the start and gradually opened to permit more ventilation toward the end of the incubation period. Follow the manufacturer's directions, or if using a homemade machine, experiment until you obtain the best results. The room where the incubator is located should be ventilated, yet not drafty. For small tabletop polystyrene incubators, keep the room temperature above 68°F (20°C).

Humidity

Humidity in the incubator varies from 83 to 88°F (28–31°C) (wet-bulb thermometer; 50 to 60 percent relative humidity [RH]), depending on the type of eggs. This level of humidity is maintained until the last 3 or 4 days before hatching, and then is increased to 90 to 95°F (32–35°C) (wet-bulb; 60 to 75 percent RH). A wet-bulb thermometer can be made by taking a regular bulb-type thermometer and placing a length of clean cotton wick (hollow cotton shoelace) over the bulb, and placing the extra wick in a small container of water, allowing air to move freely across the bulb and the wick. The higher the humidity, the closer the wet-bulb temperature is to the dry-bulb temp and vice versa.

The humidity in small incubators is provided by moisture-evaporation pans. Keep the pans full at all times. When adding water, add warm water (110°F [43°C]) so as not to reduce the incubator temperature for an extended period of time. If the humidity in the machine needs to be elevated (a higher wet-bulb-thermometer reading), increase the evaporative surface area. A sponge placed in the evaporation pan will help increase surface area and thus the moisture level. More pan surface area also helps.

During the incubation period (19 days), eggs should lose about 11 to 12 percent of their original weight through evaporation. A loss of more than this amount is detrimental. The amount of evaporation is controlled by proper humidity. The relative-humidity level for chicken eggs should be 50 to 55 percent for the first 18 days (83 to 87°F [28–30°C] wet-bulb reading) and 65 percent (89 to 90°F [31.6–32.2°C] wet-bulb reading) during the last 3 days. If humidity is too low, close the vents partway, or add to the moisture evaporation surface. If humidity is too high, permit more ventilation.

The illustration shows the approximate size of the air cell when evaporation is normal. By candling the egg at various stages of incubation, this can be used as a rough guide to control humidity in the absence of a wet-bulb thermometer. The candling procedure and candling light are covered in the section on candling, later in this chapter.

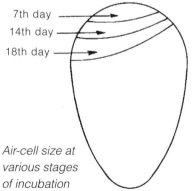

7th day
14th day
18th day

Air-cell size at various stages of incubation

Turning the Eggs

While eggs are incubating, they should be turned at least three times — or any odd number of times — a day to prevent the embryo from sticking to the shell membrane. Large machines may include automatic or mechanical turning devices. Small incubators sometimes have turning devices, but usually the eggs have to be turned by hand. When doing so, be sure to turn the eggs an odd number of times each day when they are not turned during the night. If the eggs are not turned during the night, they should be turned late in the evening and early in the morning. If eggs are set on their ends, they should be slanted at about 30 degrees. If they lie flat in the tray, they should be turned from one side to the other — in other words, 180 degrees. One method of making sure that all eggs are turned is to put an X on one side of the egg and an O on the other, in pencil, making a line between the X and the O on one side. Then turn between the X and the O alternately, following the line. When turning the eggs, make sure all the Xs or Os are on top. Discontinue turning after 18 days for chicken eggs, or 3 days prior to hatching for eggs of all other species.

Candling

Eggs are normally candled 3 days prior to hatching, at which time the infertiles are removed. Candling is done in a dark room using a special light. The illustration on page 130 shows a candling light that you can easily make at home. If the eggs are infertile, they will appear clear before the light. The fertile eggs will permit light only through the large, or air-cell, end of the egg. The rest of the egg will be black or very dark. Eggs may be candled earlier in the incubation period: At 72 hours, the early fertiles will have the typical blood vessel formation, looking much like a spider.

After the eggs hatch, leave the young birds in the incubator for approximately 12 hours, until they are dried and fluffy, before removing them. The young can survive without food and water for 72 hours, but the sooner they are put on feed and water, the better.

The next two charts, Incubation Periods for Various Species (page 130), and Incubation Troubleshooting Chart (pages 131–132), contain incubation troubleshooting information.

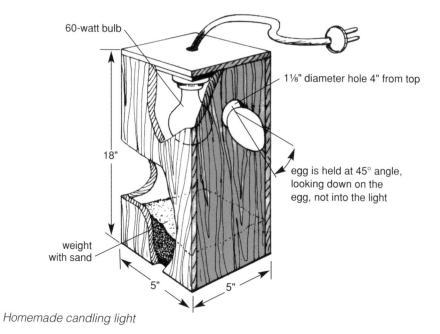

60-watt bulb

1⅛" diameter hole 4" from top

18"

egg is held at 45° angle,
looking down on the
egg, not into the light

weight
with sand

5"   5"

Homemade candling light

Incubation Periods for Various Species

SPECIES	DAYS
Chicken	21
Duck	28
Muscovy duck	33–37
Turkey	28
Goose	30–32
Canada and Egyptian Geese	35
Guinea fowl	26–28
Pheasant (Ring-Necked)	23–24
Mongolian Pheasant	24–25
Ostrich	42
Pigeon	16–20
Japanese Quail	17
Bobwhite Quail	23
Peafowl	28

Incubation Troubleshooting Chart

Symptom	Probable Cause	Suggested Remedies
Eggs clear No blood ring or embryo growth	Improper mating	8 to 10 vigorous males per 100 birds
	Eggs too old	Eggs set within 10 days after date laid
	Brooding hens too thin	Keep hens in good flesh
	Too closely confined	Provide recommended amount of floor space per bird
Eggs candling clear But showing blood or very small embryo on breaking	Incubator temperature too high	Watch incubator temperature
	Badly chilled eggs	Protect eggs against freezing temperature
	Breeding flock out of condition (frozen combs, chicken pen)	Do not set eggs from birds with frozen combs or with contagious diseases
	Low vitamin ration	Feed fish oil and alfalfa
Dead germs Embryos dying at 12–18 days	Wrong turning	Close temperature regulation
	Lack of ventilation	Plenty of fresh air in incubator room and good ventilation of machines
	Faulty rations	Feed yellow corn, milk, alfalfa meal, and fish oil
Chick fully formed But dead without pipping	Improper turning	Turn eggs 3–5 times daily
	Heredity	Select for high hatchability
	Wrong temperature	Watch incubator temperature
Eggs pipped Chick dead in shell	Low average humidity	Keep wet-bulb temperature at 85–90°F
	Low average temperature	Maintain proper temperature throughout hatch
	Excessive high temperature for short period	Guard against temperature surge
Sticky chicks Shell sticking to chick	Eggs dried down too much	Carry wet-bulb temperature at 85°F between hatches
	Low humidity at hatching time	Increase wet-bulb reading to 88–90°F when eggs start pipping

chart continues on page 132

Incubation Troubleshooting Chart (continued)

SYMPTOM	PROBABLE CAUSE	SUGGESTED REMEDIES
Sticky chicks Chicks smeared with egg contents	Low average temperature	Proper operating temperature
	Small air cell due to high average humidity	Increase ventilation and lower humidity
Rough navels	High temperature or wide temperature variations	Careful operation
	Low humidity	Proper humidity
Chicks too small	Small eggs	Set nothing under 23-oz. eggs
	Low humidity	Maintain proper humidity
	High temperature	Watch incubator temperature
Large soft-bodied chicks	Low average temperature	Proper temperature
	Poor ventilation	Adequate ventilation
Mushy chicks	Navel infection in incubator	Careful fumigation of incubator between hatches
Short down on chicks	High temperature	Proper temperature
	Low humidity	Careful moisture control
Hatching too early With bloody navels	Temperature too high	Proper control of temperature
Draggy hatch Some chicks early, but hatch slow in finishing	Temperature too high	Proper operation
Delayed hatch Eggs not starting to pip until 21st day or later	Average temperature too low	Watch temperature, check thermometers
Crippled chicks	Cross beak — heredity	Careful flock culling
	Missing eye — abnormal	Matter of chance
	Crooked toes — temp.	Watch temperature
	Wryneck — nutrition?	Not fully known
Excessively yellow	Too much formaldehyde, fumigation	Follow directions on fumigation program

Source: Embryology and Biology of Chickens, *University of Vermont.*

Natural Incubation

The first requirement for hatching eggs naturally is a good setting hen. Many of our modern-day chickens have had much of their broody behavior bred out of them. Rhode Island Reds, Plymouth Rocks, Wyandottes, and Orpingtons are usually good brooders. Don't use a bird with long spurs: It is likely to trample the chicks. Select a calm bird that will be less likely to break eggs in the event that it becomes frightened.

Provide a suitable nesting box, one that is roomy and deep, so that it will have plenty of room to change its position, turn the eggs, and be comfortable. Select a quiet spot for the nest, away from dogs or other predators and disturbances. Darken the nest area and provide feed and water close by. The bird should be free of lice or mites before setting.

Constructing a Small Incubator

You can construct a small still-air or forced-draft incubator quite easily and inexpensively. There are several ways to make them. If the incubator is to be used for demonstration purposes, make the top or one side of glass, so that the hatching process can be observed. Some excellent still-air incubators have been constructed from polystyrene coolers. Others have been made from cardboard boxes or fabricated from plywood. It may be less expensive to purchase a ready-made incubator.

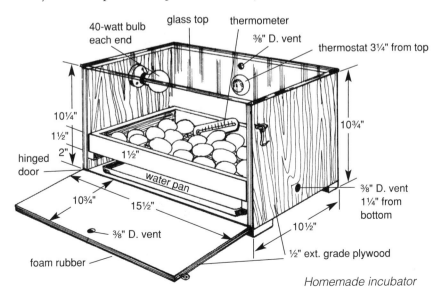

Homemade incubator

Some manufacturers offer incubator plans and kits for sale; and, of course, some companies also sell incubators. A list of some companies selling incubators and incubator parts may be found in Sources of Supplies and Equipment (see page 323).

Sex Identification

Chicks can be purchased straight-run from the hatchery. Straight-run or as-hatched chicks will be 50 percent females and 50 percent males. A few small-flock owners sometimes order straight-run, dual-purpose chicks and raise the males for broilers and roasters and keep the pullets as layers; however, it is usually less expensive to purchase meat birds for this purpose. Large commercial producers and most small-flock producers prefer sexed chicks. Thus, there is a need for sex differentiation at the hatchery.

Vent-Sexing Method

Several methods are used in the industry to sex day-old chicks. The Japanese, or vent-sexing, method has been used for many years. The procedure requires an examination of the rudimentary copulatory organs. An efficient sexer must have good eyesight, be well trained, and keep in practice to be accurate. There is great variation in the appearance of the male organs and of the female organs, making it a challenge for the inexperienced to accurately determine the sex of day-old chicks.

Sex-Link Method

Another method often used involves sex-linked inheritance. Sex-linked characters are transmitted from the female of a given mating to its sons but not to its daughters. When sex-linked genes produce recognizable characteristics in newly hatched chicks, these characters can be utilized to identify the sex of the offspring.

The Rhode Island Red male crossed with the Barred Plymouth Rock female is an excellent example of sex linkage and its use in identifying the sex of baby chicks. The female chicks will be nonbarred and the male chicks will be barred. The barring will not show on the down of the newly hatched chicks but will show up as a white spot on the back of the head.

The females will not have the spot and when mature will be predominantly black, while the males will have a white spot on the head and exhibit barring. The chicks can be sorted easily.

A second example of sex linkage is rate of feathering. Slow feathering is dominant over fast feathering, so the cross is made between a rapid-feathering male and a slow-feathering female. At hatching, the rapid-feathering female chicks from this cross will show well-developed primary and secondary wing feathers, while the slow-feathering males will have primaries and secondaries that are much shorter.

The Avian Egg

The avian egg is one of nature's marvels. We tend to think of it as a source of food, and certainly the egg is important as a food. In fact, it is probably the most nearly complete food known to man. The American Egg Board refers to it as "the incredible edible egg."

Nature, however, intended the egg to be a means of reproduction. It is actually a single reproductive cell surrounded by all the nutrients needed by the developing embryo. It provides enough residual nourishment for the baby chick to maintain itself for 72 hours after hatching.

The egg (ovum) remains a single cell until fertilized by the single cell (nucleus) of the male sperm. Fertilization occurs about 24 hours before the egg is laid, just after ovulation. At that point, it carries the full complement of chromosomes and genes. The completed single cell (zygote) rapidly divides into two cells, four, eight, sixteen, and so on, after fertilization. The hen's body temperature is 105 to 107°F (40–42°C), so cell

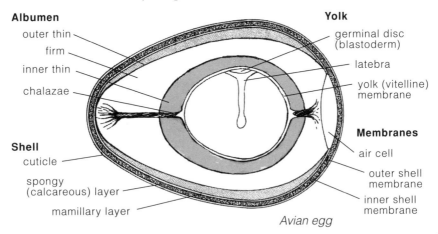

Avian egg

division continues; by the time the egg is laid, several thousand cell divisions have taken place. When the egg is laid and cools, cell division stops, to be resumed when exposed to the right conditions of moisture and heat. In nature or where permitted, the female provides these conditions when she sets on the eggs. It is done artificially with an incubator.

Most people, young or old, are fascinated by chicks hatching from eggs. Several projects using the egg as a study tool have been designed to help provide an understanding of some of the principles of reproduction. Before discussing these, it is important to look at the parts of the avian egg and their role in the process of embryo development, as well as the reproductive system of, particularly, the female, egg formation, and how fertilization takes place.

The Shell

This hard protective covering of the egg is composed primarily of calcium carbonate. The shell contains approximately seven thousand pores. The pores at the larger end of the egg are bigger and more numerous. The pores permit the exchange of gases: Carbon dioxide and moisture are given off through the shell pores, and atmospheric gases such as oxygen are taken in. When the egg is laid, a waxlike substance, called the *cuticle*, is added to the outside of the shell to protect it from penetration of dirt and disease organisms.

The Membrane

Two tough membranes are attached to the shell. They are the outer and inner shell membranes, and they prevent rapid evaporation of liquid from the egg and also prevent bacterial contamination.

The Air Cell

As the egg cools after being laid, the egg contents, which practically fill the shell at this point, contract, bringing in atmospheric air through the pores at the large end of the egg. The air cell is thus formed, providing the first air for breathing when the chick begins to pip its way out of the shell.

The Albumen

The albumen provides a liquid medium for the developing embryo. It also contains a large amount of protein necessary for embryonic growth.

The Chalazae

In a fresh egg, two white twisted cords are attached to the yolk on one end and into the outer layer of thick albumen on the other. These are called chalazae. The other ends of the chalazae are attached to the outer layer of thick albumen. The chalazae hold the yolk centered in the egg and keep it from adhering to the shell membranes and damaging the embryo.

The Yolk

The yolk contains large amounts of protein, carbohydrates, fats, vitamins, and minerals, all essential for normal growth. These substances combine with oxygen taken in through the shell pores to provide an abundant source of metabolic energy for the developing embryo. By-products of these metabolic processes are carbon dioxide and water. The carbon dioxide is given off through the shell. The water is used by the embryo to replace that lost through evaporation. Calcium absorbed from the shell and the yolk is used for the development of the skeletal system.

The Reproductive System and Fertilization

The reproductive system of the egg has two parts: the ovary and the oviduct. Unlike most female animals, the hen has only one functioning ovary, the left one.

The Ovary

The ovary is a cluster of developing yolks attached to the hen's backbone. It is located about midway up the back. The ovary contains several thousand immature ova when the female chick is hatched. As the female reaches maturity, the ova start to develop a few at a time.

The Oviduct

The oviduct is a tubelike structure lying along the backbone. It is approximately 27 inches (68.6 cm) long in a mature fowl. When the yolk is completely formed on the ovary, the follicle sac ruptures, releasing it from the ovary. The yolk then enters the oviduct entrance, or infundibulum or funnel, where fertilization takes place and the chalazae layer of albumen is added. The egg then moves through the magnum, where most of the albumen is added, and then on to the isthmus, where the shell membranes are added. Next it moves into the uterus, or shell gland, where the shell is formed around the egg. The cuticle layer is added in the shell gland also. Shell formation takes about 20 hours. From the shell gland, the egg moves into the vagina in preparation for ovipositon (lay). The egg travels down the oviduct small end first, but in the vagina, it rotates so it is laid large end first

The male has two testes located high in the body cavity along the back. When the male mates with the female, it deposits spermatozoa into

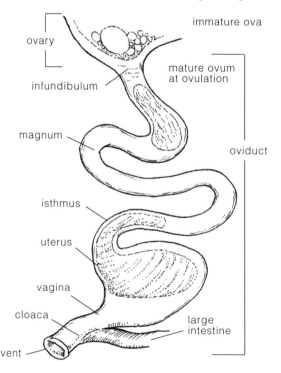

The oviduct (redrawn from Mickey A. Latur, Purdue University)

the oviduct. The sperm containing the male germ cells travel on through the oviduct, where some are stored in sperm nests in the vagina and the rest in sperm nests in the infundibulum.

The egg yolk has a tiny white spot on the surface called the *blastodisc*. The blastodisc contains the single female cell. When the yolk enters the infundibulum, a sperm penetrates the blastodisc and fertilizes it. The blastodisc then becomes a blastoderm.

Methods of Observing Embryo Development

A number of methods can be used to observe embryo development during the incubation period. Some of these are candling, the shell-window method, and the in vitro method. Egg candling was discussed earlier. It is a useful method to determine if eggs are fertile early in the incubation period, and also for observing embryo development later during incubation.

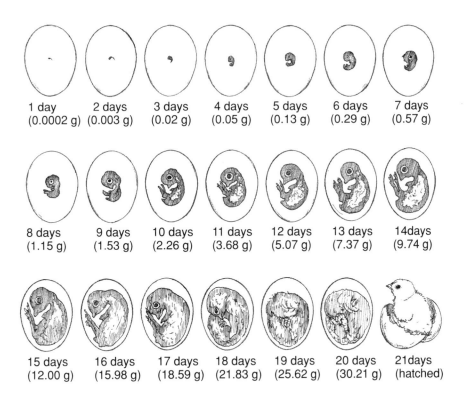

| 1 day | 2 days | 3 days | 4 days | 5 days | 6 days | 7 days |
| (0.0002 g) | (0.003 g) | (0.02 g) | (0.05 g) | (0.13 g) | (0.29 g) | (0.57 g) |

| 8 days | 9 days | 10 days | 11 days | 12 days | 13 days | 14days |
| (1.15 g) | (1.53 g) | (2.26 g) | (3.68 g) | (5.07 g) | (7.37 g) | (9.74 g) |

| 15 days | 16 days | 17 days | 18 days | 19 days | 20 days | 21days |
| (12.00 g) | (15.98 g) | (18.59 g) | (21.83 g) | (25.62 g) | (30.21 g) | (hatched) |

Development of the embryo

The Shell-Window Method

As the name suggests, the shell-window method involves cutting the large end of the egg to allow a view of the embryo. Remove the shell carefully to avoid damaging the embryo. Crack the shell and cut or peel it away, using care not to puncture the inner shell membrane. After the shell is removed, peel away the inner membrane using forceps, tweezers, or scissors. The embryo may be stuck to the inner shell membrane. If so, moisten the membrane with warm water from a small dropper to loosen it, and peel away the membrane. It is now possible to observe the embryo through the window in the shell. If covered with plastic wrap and put back into the incubator, the embryo will survive for a few hours to a few days.

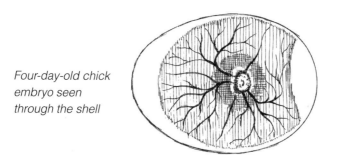

Four-day-old chick embryo seen through the shell

The In Vitro Method

The in vitro procedure provides an excellent means of observing embryo development without the interference of the shell. This method makes it possible to observe the beating heart and the early development of the circulatory system. The eye may also be seen easily after the 3 days of incubation. As the embryo slowly develops, it is easy to observe the process and day-to-day changes.

Three-day embryos are removed from the shell and placed in plastic pouches or slings. The pouch is made of clear plastic and suspended by a ring. The containers may be removed from the incubator periodically for observation and then placed back into the incubator for further development. With care, many of the in vitro embryos live for several days.

Fertile eggs must be incubated a full 72 hours before they are ready to be prepared for the in vitro. Needed equipment and directions for preparation follow. Wash your hands thoroughly before starting.

Materials and Equipment

- **Incubator** — with 3-day embryos
- **Plastic wrap** — inexpensive or generic is best; expensive wrap or microwave wrap is not porous enough to permit exchange of gases needed for healthy embryos
- **Rings** — 4-inch-diameter (10 cm) plastic pipe cut in 2-inch (5 cm) segments or 4-inch-diameter (10 cm) plastic cottage-cheese containers with part of the bottom removed
- **Rubber bands** — strong and fairly large; two per ring
- **Scissors** — used to trim the wrap
- **Receptable** — for eggshells and paper towels

1. Cut two 1-foot (30.5 cm) lengths of plastic wrap for each ring to be prepared.

2. Drape one piece of plastic wrap over a ring. Using both hands, place one of the rubber bands around the ring and the plastic wrap. Let the rubber band roll off your fingers ½ inch (1.3 cm) from the bottom of the ring and adjust the wrap to remove the wrinkles.

3. Using the fingers of both hands, slide the wrap and rubber band to within ½ inch (1.3 cm) of the top of the ring. This will permit the accumulation of enough wrap to form a pouch or cradle for the embryo inside the ring. Push down gently on the wrap to form the pouch. The bottom of the pouch should extend down in the center to about ½ inch (1.3 cm) from the bottom of the ring. Adjust if necessary. ▼

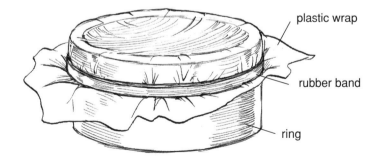

plastic wrap

rubber band

ring

Step 3

4. Remove each 3-day embryo from the incubator one at a time as you prepare the in vitro, so as to avoid chilling the embryos. Hold the egg still on its side for 25 seconds in the position you would if you were going to break it out into a frying pan for cooking. This enables the embryo to float to the top and helps prevent the rupture of blood vessels when the egg is broken into the pouch. Strike the underside of the egg on a hard surface, lower the egg as close to the bottom of the pouch as you can, and gently pry open the shell and deposit the egg contents in the pouch. What you should see on the top of the yolk is a faint circle of blood vessels about the size of a nickel. The heart is deep brown in color and shaped like a comma; it should be beating. If the yolk is clear, carefully roll or turn it over, using your fingers, and look for signs of embryo development.

5. When you have determined that you have a viable embryo, drape the second piece of plastic wrap over the ring. Place the second rubber band over the wrap and around the ring, and adjust to pull the wrap tight. Trim away the excess wrap. Keep the ring on a flat surface while preparing the in vitro. Be careful not to jiggle or tip it any more than necessary.

6. Place the in vitros in the incubator as they are completed. In vitros are not turned in the incubator. Several in vitros should be prepared, to ensure an adequate number of observations and enable comparison. The purpose of the in vitro is to make embryo development observable. It has to be handled to be seen, but be careful. Avoid jarring, tipping, or excessive cooling. One or two out of the incubator should be sufficient for most observations.

It is difficult to predict how many embryos will die and when. Many die at 6 to 8 days, yet others live 16 to 19 days.

PROSPEROUS EGG BUSINESS

Mary L. was born on a farm in 1909 and grew up around animals. However, poultry did not play a major role in her life until after she was married. She and her husband John started a farm of their own and grew broilers until John's death in 1955. The broiler business was in its infancy at that time and Mary was not making a go of it. With five children to feed and care for, she needed something that would give her a steady income. She decided to raise laying hens and developed a prosperous egg business. It was so successful that it was studied as a model business by students from other towns, states, and countries.

The farm grew; and to keep more control of the costs of rearing poultry, Mary and another local poultry farmer formed a partnership and built a feed mill, which supplied feed not only for their birds, but for many others in the area as well. Although the business passed to one of her sons and one of her partner's sons and became one of the largest egg-laying businesses on the East Coast, Mary kept a watchful eye over the business, making sure the high standards of quality were maintained, until her death at age 90.

EIGHT

MEAT-BIRD PRODUCTION

Birds grown for meat include broilers, roasters, capons, guinea fowl, game birds, turkeys, and waterfowl. Turkeys, waterfowl, and game birds seem important enough in themselves to warrant separate chapters. Thus, they are covered in chapters 9, 10, and 11, respectively.

Most meat birds, and particularly broilers, were once produced as a by-product of chick replacements for egg production. Chicks were purchased as straight run; that is, approximately 50 percent cockerels and 50 percent pullets. The cockerels were separated from the pullets at about 10 to 14 weeks of age and marketed as broilers or fryers or carried on to heavier weights as roasters. Males to be used as capons were taken out, or separated from the pullets, at about 5 or 6 weeks and surgically caponized and carried through to marketing time. Today these cockerels are destroyed at the hatchery, in most instances, because they are relatively inefficient producers of meat.

For top performance, birds to be grown for meat should be selected from those sources that have outstanding broiler birds. Today's broiler chick is a highly specialized bird, bred specifically for meat production. It is not a good egg producer but is efficient as a producer of meat. Broiler chicks are crossbreds or hybrids, and the parent stock that produces the chicks is usually of White Cornish and White Plymouth Rock breeding. White feathers are preferred in meat birds for the ease of dressing and for better carcass appearance. The Cornish White Rock crosses give chicks with white or predominantly white feathers and also produce chicks with meaty breasts and large legs, and with rapid, efficient growth characteristics. The dressed birds have excellent conformation and market appearance due to the lack of dark pinfeathers.

Breeders are continually improving their stocks. When buying a broiler chick or any chick to be used for meat production, check the growth rate and feed conversion of available stocks in your area. Contact your state Poultry Extension Specialist or county agent. It pays to start with the best chicks you can get. Chick cost is a small portion of the total cost of producing a pound of poultry meat.

The series of charts on pages 146–147 present information on broiler growth, feed consumption, and feed conversion; water-consumption data; and the costs of broiler production.

Most meat birds are sold as day-old chicks. A possible exception is the started capon, which may be purchased at 4 to 5 weeks of age from a hatchery or a dealer.

Housing and Equipment Needs

Meat birds can be grown either in confinement or on range. Most broilers are grown in confinement. For at least the first several weeks, they have to be confined to the brooder house, and because the males can be dressed as early as week 6, it usually doesn't pay to move them to range. The best growth results are obtained by rearing in confinement.

Roasters and capons are sometimes grown on range or yarded. This is satisfactory where housing is scarce and the climate is moderate, but growth and finish may be sacrificed by doing this. Meat birds usually do better when confined and receiving a full feed of a specialized ration designed for rapid growth and efficient conversion.

Floor Space

Meat birds should receive at least ½ square foot (0.05 sq m) of floor space per bird until they are 2 weeks old and 1 square foot (0.09 sq m) per bird between weeks 2 and 10. Heavy meat birds such as capons and roasters from 10 to 20 weeks of age should have 2 to 3 square feet (0.2–0.3 sq m) of floor space per bird. When meat birds are kept more than 20 weeks, they should have a minimum of 3 square feet (0.3 sq m) per bird.

Standards for Broiler Growth and Feed Consumption

	LIVE WEIGHT (POUNDS)		
WEEK	MALES	FEMALES	MIXED SEXES
1	0.36	0.34	0.35
2	0.90	0.84	0.88
3	1.72	1.57	1.65
4	2.80	2.50	2.66
5	4.05	3.53	3.79
6	5.38	4.55	4.97
7	6.71	5.52	6.11
8	7.97	6.40	7.18
9	9.09	7.43	8.17
10	10.09	8.32	9.20

*Feed conversion = pounds of feed consumed, divided by the live weight of the birds.

Approximate Daily Water Consumption of Broilers

	GALLONS WATER PER 100 BIRDS	
AGE IN WEEKS	70°F (21°C)	90°F (32°C)
0–1	0.8	0.9
1–2	1.6	2.6
2–3	2.5	5.2
3–4	3.5	7.2
4–5	4.6	9.4
5–6	5.7	11.0
6–7	6.7	12.2
7–8	7.6	12.5

Source: Mack O. North, Donald D. Bell, Commercial Chicken Production Manual, *4th ed. New York: Von Nostrand Reinhold, 1990.*

Gain over Preceding Week (mixed sexes) Pounds	Feed per Broiler for the Week (mixed sexes) Pounds	Feed to Date (mixed sexes) Pounds	Feed Conversion* to Date (mixed sexes) Pound of Gain to Pound of Feed
0.25	0.28	0.28	0.80
0.53	0.67	0.95	1.08
0.77	1.17	2.12	1.28
1.01	1.69	3.81	1.43
1.13	2.20	6.01	1.58
1.18	2.65	8.66	1.74
1.14	3.00	11.66	1.91
1.07	3.62	15.28	2.20
0.99	3.85	19.13	2.34
1.03	4.31	23.44	2.54

Estimated per-Bird Costs of Broiler Production

Item	Cost
Chick cost	$.45–.65
Brooding	.10
Feed (9.2 lb at $.10)	.92
Litter and misc.	.10
TOTAL per bird	$1.52 –1.72

Feed and Water Space

The chicks may be fed in box lids or feeder trays for the first 4 to 5 days, or chick-sized feeders may be used at the rate of 1 inch (2.5 cm) per chick. At 3 to 6 weeks, they should have 2 inches (5 cm) of feeder space per chick and at 7 to 12 weeks, 3 inches (7.6 cm). Birds to be carried on as roasters should be provided with 4 inches (10.2 cm) of feeder space per bird for the remainder of the growing period. If round feeder pans or tube feeders are used, the feeder space may be reduced by about 20 percent.

The feed trough should not be more than half full to avoid feed waste. Adjust feeders so that the lip is level with the bird's back. Adjustable hanging tube feeders are excellent for growing birds. Three tube feeders will normally accommodate one hundred birds.

Provide water at the rate of 20 inches (50.8 cm) of trough space or two 1-gallon (3.8 L) waterers per one hundred chicks up to 3 weeks of age. From week 3 they should have a 4- to 5-foot (1.2–1.5 cm) trough per one hundred chicks, or two large fountain waterers.

Housing Requirements

General housing requirements — for cleanliness, sanitation, litter, the arrangement of feeding and watering, and brooding equipment — are all much the same as those described for brooding replacement pullets in chapter 5. Roosts are not recommended for brooding meat birds. Roosting can cause such problems as crooked breastbones and breast blisters.

Lighting

Lighting meat birds is somewhat different from lighting replacement birds. Enough light should be provided the broiler chick to enable it to move about and see to eat and drink. Actually, it is desirable to keep activity at a reduced level for the most efficient feed utilization. The intensity of illumination at the level of the chick should be about ½ foot-candle. It is sometimes difficult to keep this level of lighting in housing with windows. Birds that receive a higher intensity are more prone to feather picking and cannibalism, increased activity, and possible piling and smothering. If you are using an incandescent bulb, then 1 bulb watt per 8 square feet (0.7 sq m) of floor space will normally provide about

½-foot-candle intensity. This is when the bulb has a reflector and is 7 to 8 feet (2.1–2.4 m) above the floor.

Many broiler producers use continuous light for the first 2 or 3 days, and then switch to 23 hours of light to 1 hour of dark (23L:1D) for the remainder of the growing period. This may be somewhat hazardous. In the event of a power failure, the birds may panic and pile. However, providing 23L:1D for the first 2 or 3 days and then 20L:4D will provide optimum growth and immunity and still maintain good feed efficiency.

Management Systems for Meat Birds

Although some research is being done to develop cage equipment for broilers and other meat birds, the recommended management system is still the conventional floor or litter system. Cage growing offers several advantages, but thus far the disadvantages outweigh them. More breast blisters, leg and foot problems, crooked keels, poor growth, and feed inefficiency are problems that may result with cage management systems of growing meat birds.

The windowed house or controlled-environment house is used for meat birds, depending upon the climate, the size of the flock, and other factors. Ventilation and ventilation systems are the same as those for other types of growing houses. In most areas, concrete floors are required for proper cleaning and sanitation and to avoid disease and parasite problems. The cleaning and sanitation requirements are the same for meat birds as for other growing stock.

Disease Control

Disease prevention is vitally important to a broiler growing program, and it is important to purchase birds that come from flocks free of *Salmonella enteritidis*, *Salmonella pullorum*, typhoid, *Mycoplasma gallisepticum*, and *Mycoplasma synoviae*. Unlike birds grown for egg production, or meat birds such as roasters and capons, broilers typically have only a 6- to 8-week growing period. If the birds become sick in the middle of the growing period, the time in which the disease can be treated and brought under control is relatively limited, and the overall flock results can be severely affected. It is, therefore, important that a disease-control program for broilers be one of prevention rather than treatment.

Vaccination

Vaccination recommendations for meat birds vary with the area. It is good to have the chicks vaccinated for Marek's disease, especially if it has been prevalent in your area. Marek's vaccine is usually administered at the hatchery. Some broiler growers do not vaccinate for other diseases because they practice isolation and good sanitation and believe that these programs will prevent most disease outbreaks.

Before establishing a vaccination program for meat birds, be sure to check with poultry pathologists or other people who are knowledgeable about the possible disease exposure situation in your area. When broilers or other meat birds are vaccinated for something other than Marek's disease, it is normally Newcastle disease or infectious bronchitis.

Coccidiosis

It is vitally important to prevent outbreaks of coccidiosis in meat birds. Such outbreaks normally occur early in the chick's life and can damage the intestinal lining. This prevents the normal absorption of food materials in the intestinal tract. Outbreaks of coccidiosis can cause poor growth and feed conversion, or — in severe cases — mortality and morbidity. Coccidiostats can be used in the water or feeds, either to reduce the infection of coccidiosis or to completely suppress it.

Because the growing period of broilers is so short, and so much emphasis is placed on fast, efficient growth, the usual procedure is to control coccidiosis during the growing period. Better growth results are obtained if infection is prevented. Most roaster and capon growers use a coccidiostat in the feed throughout the growing period.

Even if the birds are on a preventive level of coccidiostat, at times there are outbreaks of the disease due to excessively wet litter or other factors. The method of treating these outbreaks is covered in the section on flock health, chapter 13.

Feeding Programs for Meat Birds

The goal of the feeding program for meat birds is to get as much growth as quickly as possible. This is unlike feeding replacement pullets, where delaying sexual maturity is important. Meat birds must have access to full feed at all times. A number of small-flock owners permit capons and roasters to yard or range. Although overall performance may be somewhat less, excellent birds can be grown this way. The cost may be somewhat less, also, because the chickens forage for a part of their food intake.

It is important to feed meat birds a ration designed to give rapid growth and good feed conversion. Start them on a broiler feeding program. The complete feeding program includes two or possibly three different rations. Some companies recommend a program that begins with a broiler starter the first 3 weeks. The starter contains about 23 percent protein and a coccidiostat. This feed is replaced with a 20 percent broiler grower from 3 weeks up to market or up to 7 to 10 days prior to marketing. At that time, they are shifted to a final feed of 18 percent, containing no coccidiostat. If you use this feeding program, you must check the birds very carefully for coccidiosis if they are to be held beyond the 10-day period. It may become necessary to treat for coccidiosis if you hold them more than 5 days prior to dressing or marketing.

If birds are to be held over as light roasters for 4 to 5 weeks beyond the broiler period, they can be fed an 18 to 20 percent broiler finisher diet. This may be fed up to 2 to 5 days prior to dressing or marketing, depending upon the type of coccidiostat used. Some coccidiostats must be withdrawn several days prior to slaughter to avoid carcass residues.

If birds are fed as capons or heavy roasters, they should be fed a lower-protein and higher-energy diet from 9 or 10 weeks to marketing. The protein level should be approximately 16 percent. Some feed-company programs call for feeding a roaster finisher from 6 weeks to 2 days before marketing, at which time a feed containing no coccidiostat is used for the remainder of the period.

Capon Production

Male chickens castrated or caponized to make them fatten more readily are called *capons*. Capons are not only more tender than cockerels but bring a much better market price as well. They are normally grown to be marketed during the Thanksgiving and Christmas holiday seasons, at which time they bring premium prices.

Up to possibly 14 weeks of age, the capon and the cockerel weigh about the same. Afterward, the capons gain weight more rapidly than the cockerels. When cockerels show spur and comb development, they are called stags, and their flesh toughens. Toughness of the flesh does not occur in capons. When cockerels are a year old, and are classed as old cocks, they bring a low price and their flesh is tough.

Unlike the cockerel, the capon has a quiet disposition, is docile, and seldom crows. After caponization, the comb and wattles cease growing, the head looks pale and small, and the hackle, tail, and saddle feathers grow to be unusually long.

Selecting a Breed for Caponizing

Broiler stocks with white plumage and yellow skin are preferred for meat-bird production. White Plymouth Rocks, New Hampshires, and crosses of these breeds, or those crosses involving White Cornish and White Rock Crosses, are generally used. Caponizing small birds, such as Leghorns or the small representatives of the American breeds, just does not pay.

When to Caponize

Because capons are in the biggest demand during the Thanksgiving-to-March period and take 14 to 18 weeks to grow and finish properly, the best time to caponize is in late summer or fall.

With our faster-growing broiler strains of birds, it is now desirable to caponize at approximately 12 to 20 days of age, before they become too large and the operation is more difficult. Some hatcheries or dealers specialize in selling 4-week-old capons that have been caponized as early as 10 days of age.

The Caponizing Operation

Withhold feed and water from the birds for 24 hours before the operation. The bird's intestines should be almost empty to prevent them from obstructing the operation. When the birds are removed from feed and water, keep them in wire or slat-bottom coops to prevent them from eating litter or other materials.

Good light is mandatory during the operation. Direct sunlight is best; a strong electric light with a reflector can, however, serve indoors.

A barrel or box can be used as an operating table. When a large number of birds are to be caponized, a table of convenient height is recommended.

During the operation, the birds can be restrained by a second operator who holds each bird outstretched on its side. A caponizer can do the operation single-handedly by restraining the birds with weighted cords tied to the legs and wings. These cords should be about 2 feet (0.6 m) long and weighted with 1-pound (0.5 kg) weights. The cockerel's legs are securely tied by the use of a half hitch in one cord. Both wings are held together near the shoulder joint with a half hitch in the other cord. The weights are hung over the edge of the operating table, and the bird is stretched out. Arrange the cords so that it is easy to adjust the bird and turn it over without disturbing the weights.

Technique

The operation can be done with one incision or two incisions. Most operators find it easier to remove the upper, or nearer testicle, then turn the bird over and make a second incision on the other side of the body to remove the other testicle. If both testicles are removed through just one incision, it is better to remove the lower one first; otherwise, bleeding from the upper may obscure the lower.

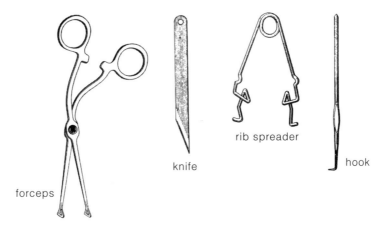

forceps   knife   rib spreader   hook

The instruments needed for the operation are a sharp knife, a rib spreader, a sharp pointed hook, and the testicle removers or forceps. The operation should be performed as quickly as possible, to reduce stress on the bird.

It is essential not to rupture the artery that runs just behind the testicles, or the chicken will bleed to death in seconds. If feed and water have been withheld, the artery usually presents no problem.

1. Moisten and remove the feathers from a small area over the last two ribs, just in front of the thigh; with one hand, slide the skin flesh down toward the thigh and hold it in that position. This will prevent cutting into the muscle and causing excessive bleeding.

2. Make the cut between the last two ribs. The incision should be lengthened in each direction until it is about ¾ to 1 inch (2–2.5 cm) long. ▼

Step 2

3. Insert the spreader into the incision to spring the ribs apart. The intestines will be visible beneath a thin membrane.

4. Tear open the membrane with the hook to expose the upper testicle. It is usually yellow in color but may be dark. It is about the size and shape of a navy bean and is located close to the backbone and just below the front end of the kidney. If the intestines are empty, once they are pushed aside it is quite easy to see the lower testicle. It is in the same position as the upper one, but on the other side of the backbone.

5. Use the forceps to grasp the testicle. Be careful to avoid the artery at this point. ▶

Step 5

6. Remove the entire testicle with a slow twisting motion, tearing it away from the spermatic cord to which it is attached.

7. Unless you removed both testicles with the same incision, turn the bird over and perform the same operation on the other side. It is not necessary to stitch the incision, since when the skin and flesh are released, they will slide back over the incision and cover it.

The caponizing tools should be rinsed, after each bird is caponized, in a solution consisting of 10 drops of chlorine bleach per quart (1 L) of water.

Slips

Frequently, when the caponizing operation is performed for the first time, a number of birds will become what are known as *slips*. A slip is neither a cockerel nor a capon; it develops when a part of the testicle is not removed during the operation. This small piece of testicle often grows to a considerable size.

Slips have the same restless disposition as cockerels and they grow and fatten little or no better than a cockerel. Thus, slips are not as highly valued for meat purposes and do not bring as good a price if sold. Inexperienced operators may expect as much as 50 percent in slips; experienced operators can expect about 5 percent or less.

Slips develop red heads and wattles, whereas a true capon has a pale head and wattles. They can be finished early and marketed as broilers around 8 to 9 weeks of age. They may be kept for 12 to 13 weeks and sold as roasters. Older birds develop tougher meat.

Losses in Caponizing

Occasionally, even the best operators will kill some birds during the caponizing operation. However, this loss seldom exceeds 5 percent. Inexperienced caponizers frequently kill several birds, but gradually losses are reduced as operators gain experience. Any birds killed during the operation can be dressed and eaten.

Speed develops with experience. An experienced caponizer can do as many as four birds per minute. It is wise for the beginner to practice on slaughtered birds to acquire some skill before attempting to caponize a live bird. Caponizing is somewhat of a stress on the birds, so some commercial

caponizers inject them with an antibiotic solution just prior to the operation. Some capon growers administer the antibiotic solution in the drinking water for 2 or 3 days, beginning 4 or 5 days prior to the operation.
Antibiotics should not be given by way of the drinking water within 2 days
of the operation. Some antibiotics tend to cause the intestines to balloon,
thus increasing the possibility of damage to the intestine and making it
more difficult to spot the testicle.

Care and Feeding of Caponized Birds

Separate caponized birds from other chickens, and keep them separated during the growing period. Air puffs or wind puffs may develop after
the operation. Wind puffs are caused by air that gathers under the skin.
Release this air by pricking the skin with a needle or knife and pressing it
out. Usually, within 10 days after the operation, the incision is fully healed.

Breast Blisters

Breast blisters are often a problem with capons. They are likely to appear
when the capons are about half grown, and the incidence may increase as
the birds become heavier. Birds may develop breast blisters from roosts,
especially those with narrow perches, or from resting on board, concrete,
or wire floors. Don't let the litter become caked or excessively wet; keep it
soft, loose, and dry. Roasters or capons should not be permitted to roost.

Leg weaknesses also contribute to the breast-blister problem. This can
be serious because it detracts from the dressed appearance of the birds.
Birds that do have a weak-leg problem exhibit a tendency to sit down
much of the time. The keel bone is constantly in contact with the litter
and the incidence of blisters is increased, especially if litter conditions are
poor. Other factors that affect the incidence of breast blisters include the
type of equipment used, overcrowding, genetics, and loading and handling methods. Birds can receive bruises and skin irritations by bumping
into equipment or by being pushed and shoved or mishandled.

Feed

During the growing period, supply the birds with a growing feed containing approximately 17 percent protein. The growing stage is important, because during this period leg problems may occur. Capons, if
pushed too hard for rapid growth during the early growing period, may

Approximate per-Bird Costs
of Producing Capons and Roasters

ITEM	COST
Chicks	$.35–.85
Brooding, litter, medication, misc.	.20
Feed (24–38 lb [10.9–17.2 kg]) @ $.12	2.88–4.56
Surgical caponizing	0.00–.65
TOTAL	$3.43–6.26

Note: *Feed cost will vary with the age at which birds are slaughtered. Those slaughtered as light roasters will consume less feed than capons. Roasters are not normally caponized, thus there is no caponizing charge. Chick costs will vary with source, size of order, whether sexed or straight run, and other factors.*

have insufficient bone development to support their body weight. Most cockerels that are to be used for capons are actually broiler males. This stock has been selected for rapid growth and males should reach 4 to 5 pounds (1.8–2.3 kg) in body weight at 7 weeks of age. They are normally marketed at these weights. When these birds are used for capon production, it is expected that they will grow to about 9 or 10 pounds (4.1–4.5 kg) live weight in approximately 12 to 14 weeks. Some of the birds will not have strong enough legs to support these heavier weights if growth is too rapid early in the growing period. Keep this in mind when selecting a feeding program for capons. Some growers restrict feed, and thus the growth rate, from about 6 to 10 weeks of age to prevent this problem in light roasters. Feeds specially formulated for capons take this problem into consideration. Broiler rations, however, are high-energy feeds designed for maximum growth and may create the leg problems if used throughout the growing period. For heavy roasters, it is suggested that a lower-energy diet be used from approximately 6 to 10 weeks.

Rearing

Capons can be grown in confinement with a yard provided or on range. Downgrading often seems more common in the case of confinement-reared birds. The reason is that confined birds are usually pushed to higher weights more quickly. There is also a greater possibility of crowding and injuries when birds are confined.

Range birds, however, seem to have higher mortality. A lot of this is probably due to predator losses. Provide shelters for birds on range. Wire floors and perches are commonly used, but these may cause breast blisters. Cover wire floors and perches, and preferably use litter on the floor of the shelters. Shade is important for birds on range. This is especially true during the hot-weather months or in warm-weather climates. To be most valuable, the range should be clean and provide a short, succulent forage. If the range is good, you can substantially reduce feed costs. Provide birds on range with range feeders that prevent the feed from being blown away or getting wet and moldy. Adequate feeder and watering space is important regardless of whether the birds are reared on range or in confinement. Approximately 4 inches (10.2 cm) of feeder space and 1 inch (2.5 cm) of watering space per bird are required.

Finishing Diet

About 2 weeks prior to slaughter, capons should be put on a finishing diet to get them in good marketing condition in terms of fat covering, fleshing, and pigmentation. If pushed too early with finishing diets, the birds may become overly fat, go off their legs, and develop breast blisters. It doesn't pay to keep capons beyond 20 weeks for efficient feed utilization.

NINE

TURKEY PRODUCTION

Before you begin the production of turkeys, even on a small scale, you should be aware of the costs involved. The turkey poult costs substantially more than does a chick and, of course, turkeys consume considerably more feed than do growing chicks. Thus, the cost of producing turkeys is considerably higher. The chart on page 160 presents the costs involved in raising a small flock of turkeys.

Getting Started

The turkey enterprise can be started in one of several ways. You can purchase hatching eggs and incubate them or purchase day-old poults. You may also buy breeding stock, but this is an expensive way to begin. If hatching eggs are purchased, you will need equipment for incubating; if breeding stock is purchased, it is expensive not only to buy the birds but also to maintain them. Normally, most small flocks of turkeys are purchased as poults from a hatchery or an agricultural supply store.

Varieties

There are several varieties of turkeys, but those that are most important commercially are the Broad-Breasted Bronze, the Broad-Breasted White, and the Beltsville Small White. Until recently, the Bronze was the most popular variety. Most of the commercial turkey breeders now offer strains of white turkeys that may attain body weights of 18 to 20 pounds (8.2–9.1 kg) for hens and 24 to 26 pounds (10.9–11.8 kg) for toms at about 15 weeks of age.

Estimated per-Bird Costs of Raising Heavy Roaster Turkeys*

ITEM	COST
Poult	$1.75–4.00
Feed (80 lbs [36.3 kg]) @ $.12–.15	9.60–12.00
Brooding, litter, medication, misc.	.25
TOTAL	$11.60–16.25

Costs will vary with number of poults and volume of feed purchased, price of feed, as well as sources and other factors.

Broad-Breasted Bronze

The Bronze has good growth rate, conformation, and feed conversion, and about all of the qualities demanded by the turkey industry. Its basic plumage color is black, and it has dark-colored pinfeathers — a distinct disadvantage that has led to its replacement by the Broad-Breasted White.

Broad-Breasted White

The Broad-Breasted White was developed in the early 1950s from crosses of the Broad-Breasted Bronze and the White Holland varieties. In some areas of the country, it is now difficult to find the Broad-Breasted Bronze, because the current demand is so great for the Whites.

Beltsville Small White

The Beltsville Small White looks very much like the Large White from the standpoint of color and body conformation. It is a much smaller turkey that was developed at the Beltsville Agricultural Research Center in Maryland. Its body weight has tended to increase in recent years through selection by breeders. This bird may no longer be available in many areas.

Properly managed and fed, the Beltsvilles reach good market condition in 15 to 16 weeks and make excellent turkey broilers. They also make good medium roasters if held to 21 to 24 weeks. They are not as efficient converters of feed as the Large Whites but usually cost somewhat less as poults. Their main disadvantage is the difference in feed and time required to reach the same weights as the large turkeys.

Broad-Breasted Bronze   Broad-Breasted White

Beltsville Small White

Buying Poults

When buying poults, you should select a strain that is known to yield good results. As is the case with chickens, poults should originate from sources that are U.S. pullorum and typhoid clean, and preferably from breeder flocks having no history of sinusitis, or air-sac infection. Consult your state poultry specialist, your county agricultural Extension agent, or some other knowledgeable person for advice on sources of good turkey poults in your area.

Place your order for poults well in advance of the delivery date, so as to be sure to get the quality of stock you want. Birds should be ordered to arrive to permit 12 to 20 weeks for growing the birds to average market weights. It is sometimes difficult to obtain small lots of turkeys delivered from the hatchery. However, it may be possible to pick them up at the hatchery or, perhaps, have a nearby producer order a few extra poults for you. Poults that are shipped express are subject to chilling and overheating. Buy the poults as close to home as possible.

Housing Requirements

Small flocks of turkeys are normally started in the warm months of the year, so housing doesn't have to be fancy. However, the brooder house should be a well-constructed building that is easily ventilated. If a small building is not available, you can build a pen within a large building. It should have good floors that can be readily cleaned. The insulation required will depend upon the climatic conditions as well as the time of year the poults are to be brooded. A well-insulated building can save energy and brooding costs. Young turkeys must be kept warm and dry, so a well-insulated and -ventilated house is important in cool climates, particularly with large flocks.

Confinement Rearing

Confinement rearing of turkeys is most commonly used by the smaller grower. Where predators or adverse weather conditions are likely, or where there is limited range area, confinement rearing is preferred. Some producers yard their birds within fenced-in areas using the brooder house as the shelter.

Porches

Sunporches were once popular with turkey growers, and some producers still use them. The porches are attached to the brooder house or shelter. The floor is made of either slats or wire, and is often elevated to provide a space underneath for the accumulation of droppings and easy access for cleaning. Porches are fenced in on the top and sides. Coarse-mesh wire is used on the top to avoid snow loading in those areas where that may be a problem.

Porches have several advantages, in that they help acclimate the birds to the weather conditions prior to going on range. This helps reduce stampeding on the range during wind- or rainstorms. Porches also acclimate them to changes in temperature to which they are exposed on range.

Heavy-breed turkeys today are not often grown on wire or slatted porches. Breast blisters and foot and leg problems tend to develop, particularly on wire. Light-breed turkeys can still be grown on porches. A modification of the porch approach is the use of paved yards or those with a gravel or stone surface.

Ventilation

The brooder house should have windows that slide down or tilt in from the top. These are best for ventilation. Windows should be located in the front and back of the house. Usually, 1 square foot (0.09 sq m) of window area to each 10 square feet (0.9 sq m) of floor area is adequate.

Large turkey production units sometimes use windowless, controlled-environment houses. These buildings are thoroughly insulated to provide maximum heat efficiency during brooding and comfortable, well-ventilated conditions for the birds.

Ventilation systems for turkey housing are designed essentially the same as for chicken houses. Many of them use fan ventilation with the same type of inlet but a larger ventilating capacity. Fan systems are designed to move from ¾ to 1 cubic foot (0.02–0.03 cu m) per minute at ⅛-inch (0.3 cm) static pressure per pound of turkey expected at maturity. If you plan to go into the turkey business on a commercial scale, contact your state Extension Poultry Specialist for help in designing the facilities.

Preparing the Brooder House

Between broods, clean the brooder house well. This means completely removing all the old litter and thoroughly washing the floors, sidewalls, ceilings, and equipment. Disinfect the building and equipment with a material such as cresilic acid, or one of the other disinfectants discussed in chapter 6. Some of these materials can cause foot burn or eye injury. To prevent disinfectant injury to the poults, make sure that the house has an opportunity to dry for 1 week or more prior to the time the poults are put into it.

House and Sunporch System Suitable for a Small Flock

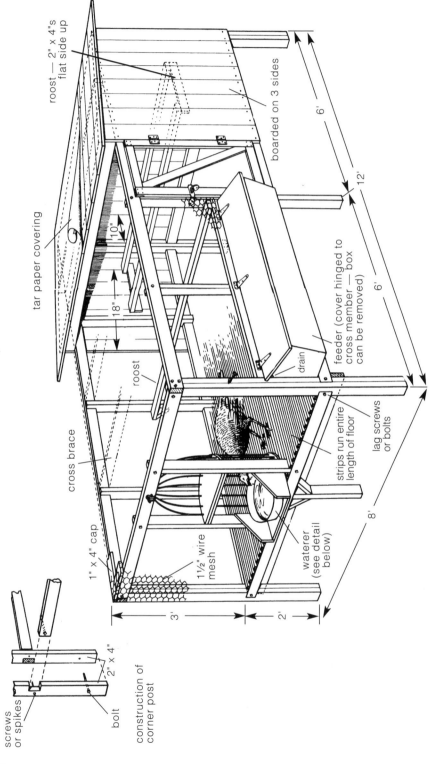

roost — 2" x 4"s flat side up

tar paper covering

boarded on 3 sides

6'

12'

10"

18"

6'

roost

cross brace

feeder (cover hinged to cross member — box can be removed)

drain

1" x 4" cap

strips run entire length of floor

lag screws or bolts

8'

1½" wire mesh

waterer (see detail below)

3'

2'

screws or spikes

2" x 4"

bolt

construction of corner post

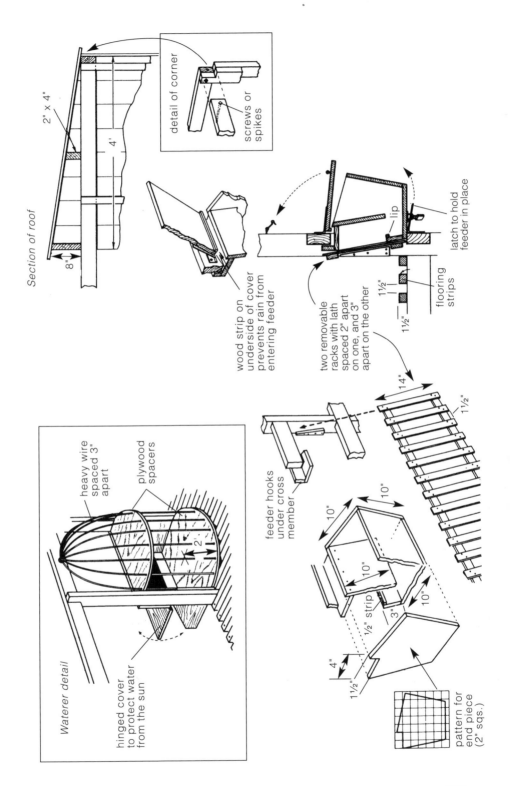

Section of roof

2" x 4"

4'

8"

detail of corner

screws or spikes

wood strip on underside of cover prevents rain from entering feeder

lip

latch to hold feeder in place

1½"

1½"

flooring strips

two removable racks with lath spaced 2" apart on one, and 3" apart on the other

14"

1½"

Waterer detail

heavy wire spaced 3" apart

plywood spacers

2'

hinged cover to protect water from the sun

feeder hooks under cross member

10"

10"

10"

10"

½" strip

3"

10"

4"

1½"

pattern for end piece (2" sqs.)

After the building is thoroughly dry, put 3 to 4 inches (7.6–10.2 cm) of litter on the floor. Some of the common litter materials suitable for use are wood shavings, sugarcane, ground corncobs, peat moss, and vermiculite. The litter may be covered or uncovered. Some producers cover it with paper to prevent litter eating for the first week. If it is covered, use a rough paper, as a slippery surface can cause leg weakness and crooked feet. The litter should be evenly distributed over the floor and free from mold and dust. Coarse litter material can also contribute to leg disorders.

In small brooder houses, corners should be rounded with small-mesh wire to prevent piling. Turkeys may pile when frightened or when the house is drafty or the floor temperature is low.

Floor Space

Use 1 square foot (0.1 sq m) of floor space per poult up to 8 weeks of age. From 8 to 12 weeks, the floor space should be increased to 2 square feet (0.18 sq m) per poult, and from 12 to 16 weeks, 2½ square feet (0.23 sq m) should be the minimum allowance. Mixed sexes kept in confinement during the entire growing period should receive 4 square feet (0.37 sq m) per bird. If the flock is all toms, provide 5 square feet (0.46 sq m) or if all hens, 3 square feet (0.28 sq m). For light turkeys or for turkeys housed in controlled-environment housing, floor-space requirements may be somewhat less than the above recommendations.

Brooders

Several types of brooders are suitable for brooding poults. Gas or electric is probably the best suited for small-flock situations. If you use hover brooders, allow approximately 12 to 13 square inches (77.4–83.9 sq cm) of hover space per poult started. A brooder that is rated for 250 chicks will be adequate for only about 125 turkeys. When you have hover brooding, use a 7½-watt lightbulb to attract the poults to the heat source. Equip each brooder with a thermometer that is easily read. Infrared brooders are satisfactory for small turkey flocks.

If you use infrared bulbs for brooding, provide two or three 250-watt bulbs per 100 poults. Hang infrared lamps 18 inches (45.7 cm) from the surface of the litter initially. They can be raised 2 inches (5.1 cm) per week until they're 24 inches (61 cm) above the litter. The brooder ring or guard should be 8 to 10 feet (2.4–3.1 m) in diameter. Take care to avoid splashing water on infrared bulbs, as this may cause them to break. Ideally, the temperature outside the heat source of the hover area should be approximately 70°F (21°C) to provide maximum comfort for the poults.

Brooder Guards

Use brooder guards to confine the birds to the heat source and to prevent drafts on the poults.

Management of the brooder guard varies, depending upon the design of the house and the climatic or seasonal conditions. Where noninsulated housing is used during fairly cool weather, an 18-inch (45.7 cm) brooder guard for each stove is recommended. For warm-weather brooding, a 12-inch (30.5 cm) brooder guard is satisfactory. Locate the brooder guard 2 to 3 feet (61–91.5 cm) from the edge of the hover and gradually move it out to a distance of 3 to 4 feet (0.9–1.2 m). After 7 to 10 days, you can remove it. At this time, the poults are normally allowed free access to the brooder house. If poults are brooded early in the season or in cold climates, and the brooder house is quite large, it may be necessary to partition off part of the house at the start and enlarge the area as necessary.

Feeding Equipment

You can supply the first feed for the poults on new egg filler flats, on chick-box lids, in feeder pans, or in small chick feeders. Turkeys sometimes have a visual problem and have difficulty finding the feed and water. Starve-outs or dehydrated poults may result. Bright-colored marbles placed on top of the feed or in the water containers will often attract the poults to the feed and water. Oatmeal or fine granite grit sprinkled very lightly over the feed, once or twice a day for the first 3 days, may also help get them eating.

Unless the litter is covered with a paper, don't fill the feeder so full that it will overflow. This can lead to the practice of eating litter. For poults from 7 days to 4 weeks old, use small feeders at the rate of 2 inches (5.1 cm) of feeder space per bird, with a 2- to 3-inch (5.1–7.6 cm) depth. From 4 to 8 weeks, the poults should be furnished with large feeders about

5 inches (12.7 cm) deep at the rate of 4 inches (10.2 cm) of feeder space per bird. From 8 weeks to market, allow about 6 inches (15.2 cm) of feeder space per bird, with about 8 inches (20.3 cm) of depth. If you use hanging tube feeders, or other round feeders, you can reduce the feeder space by about 25 percent. The number of inches of tube or round feeding space available can be determined by multiplying the diameter of the feeder pan by 3½. In figuring feeder space, remember to multiply the trough length by 2 if the poults are able to use both sides of the feeder. Thus, a 4-foot (1.2 m) trough feeder actually provides 8 linear feet (2.4 m) of feeder space.

Water Space

Turkeys require access to fresh water at all times. Water consumption can vary according to age and weight of bird, feed intake, and ambient temperature. (See page 173 for a chart listing estimates of water consumption for average-growing turkeys.) Be sure that all birds have access to waterers. Poults can be started on either glass fountain waterers or automatic waterers. From 1 day to 2 weeks, the poults should have about 24 linear inches (61 cm) of waterer space or access to two 1- or 2-gallon (3.8 or 7.6 L) fountains per one hundred poults; from 2 to 8 weeks, 48 linear inches (121.9 cm), or at least two 5-gallon (18.9 L) water founts per one hundred poults; from 8 weeks to market, about 0.75 to 1 linear inch (1.9–2.5 cm) of waterer space per bird is recommended. Add additional waterers in hot weather. Make changes of equipment — both feeders and waterers — gradually, so as not to discourage feed or water consumption.

Brooding the Poults

When the poults arrive, have the brooder house completely ready for them. Have the feeding equipment arranged and the brooder guards set up (see page 167). The brooders should have been in operation for approximately 24 hours to get them regulated and to get the building warm.

If hover brooders are used, the temperature should be 100°F (38°C) for White birds, 95°F (35°C) for Bronze. This temperature reading is taken under the hover approximately 7 inches (17.8 cm) from the outer edge and 2 inches (5.1 cm) above the litter, or at the height of the poults' backs. Be sure to check the accuracy of the thermometer you are using before the poults arrive. Reduce the hover temperature 5 degrees F (3 degrees C) weekly until it registers 70 or 75°F (21 or 24°C), or until it equals the outside temperature. If the weather is warm during the brooding period, you may shut down the heat during the day after the first week. Heat during the evening hours will be required for a longer period. Normally, little or no heat is required after 6 weeks, depending upon the time of year, the weather conditions, and the housing. Watch the poults as a guide when checking or adjusting temperatures (see page 83).

Important Reminder

Feed and water the poults as soon as possible after hatching. As each poult is taken from the box, dip its beak in water to help it learn to drink. After it has had water, place it in the feed and dip its beak into it. This will help get it started on feed and water.

Lighting

High light intensity should be provided the poults for the first 2 weeks of brooding in all types of houses to prevent starve-outs. A minimum of 2 foot-candles of light is required by turkey poults for growth. However, as much as 10 to 15 foot-candles should be used day and night. This will require 200-watt bulbs, spaced 10 feet (3 m) apart. Install a small 7½- to 15-watt attraction lightbulb under each brooder hover. After the first 2 weeks in window houses, use dim lights at approximately ½ foot-candle during the night hours. The dim lights will help prevent or discourage piling and stampeding. Turkeys are refractory to light, meaning that about a tenfold difference in brightness between the lowest level and the highest level will cause the bird to perceive the lowest level as "darkness" and the highest level as "daylight."

From 2 weeks of age to market, the birds grown in confinement in windowless houses should be given about 2 to 3 foot-candles of light on a 16-hour light to 8-hour dark schedule. Birds grown outside or in window houses can be provided with natural daylight supplemented with artificial lighting to maintain about a 16-hour day length.

Roosts

Roosts are seldom used for brooding turkeys, though they may help prevent piling at night. Roosts do not normally cause breast-blister problems with turkeys. If roosts are used in the brooder house, they may be of the stepladder type, allowing about 3 linear inches (7.6 cm) of perch space per bird at the start. The perches are made of 2-inch-round (5.1 cm) poles or 2 x 2 or 2 x 3 material, with the lowest perch located about 12 inches (30.5 cm) above the litter. Each succeeding perch is 4 to 6 inches (10.2–15.2 cm) higher. Perch edges should be beveled or rounded to prevent injury to the breasts. Each bird should have 6 linear inches (15.2 cm) of roosting space by the end of the brooding period. Screen the bottom and the ends of the roosts to prevent poults from gaining access to droppings. Birds usually begin to use the roosts at about 4 to 5 weeks of age.

In Confinement

If birds are grown in complete confinement, roosts usually are not used. Since the birds bed down on the floor, keep the litter in good condition to prevent breast blisters, soiled and matted feathers, and off-colored skin blemishes on the breast. Good litter conditions also help maintain a sanitary environment and prevent disease problems.

On the Range

Roosts used on range should be constructed of 2-inch (5.1 cm) poles or 2x4 material laid flat with rounded edges. Perches should be spaced 24 inches (61 cm) apart and located approximately 3 feet (0.9 m) off the ground. If placed in a house or shelter, roosts can be slanted to conserve space. For outdoor roosting racks, all roost perches should be built on the same level. This type of roost should be built of fairly heavy material to prevent breaking when the weight of the birds is concentrated in a small area. Ten to 15 inches (25.4–38.1 cm) of perch space per bird is required for the large birds, and 10 to 12 inches (25.4–30.5 cm) for small birds. Remember, the purpose of the roost is to provide an area for the birds to congregate and rest at night and to keep them off the ground. Range roosts are designed to be portable. They can be moved as needed to prevent the development of muddy spots and contaminated areas.

Feeding

One of the soundest pieces of advice that can be given to a turkey grower is to select a good brand of feed and follow the manufacturer's recommendations for feeding.

Two basic feeding programs for growing turkeys are available. One is the all-mash system and the second is a protein-supplement-plus-grain system. Nutrient requirements of turkey poults vary with age. As birds become older, protein, vitamin, and mineral requirements decrease and energy requirements increase.

Feed an insoluble grit such as granite grit the first 8 to 10 weeks. Whenever grains are included in the diet, or when the birds are on range, feed insoluble grit to enable the birds to grind and utilize the grains and fibrous materials.

Recommendations for feeding vary with the breeder and the type of bird you are rearing, male or female, heavy or light. Keep in mind that the number of feeds offered varies considerably among feed companies. One of the simpler types of feeding programs recommends a 28 percent starting diet, with a change to a 21 percent growing diet, and finishing with a 16 percent protein diet. Other companies offer and recommend five or six different diets during the growing period. Again, buy feed from a reputable company and follow its recommendations. Under most circumstances, the diets should include a coccidiostat.

Keep feed and water before the birds continuously. Pelleted feeds can be fed after the first 4 weeks. Most growers feed a nutritionally complete starting mash; however, you can use green feed for small flocks where labor requirements are of little concern. Tender alfalfa, white Dutch clover, young tender grass, or green grain sprouts, all chopped into short lengths and fed once or twice daily, are good for the poults. Wilted or dry roughage feeds should not be permitted to remain before the poults, as they can cause impacted or pendulous crops.

Starter Diets

Turkey starter diets can be purchased ready-mixed. They can be home- or custom-mixed according to recommended formulas, or a concentrate can be mixed with ingredients such as ground corn and soybean meal. For the most part, small producers will find it best to feed a ready-mixed complete starter diet.

Growing Diets

Growing diets, fed to the poults from 8 weeks to maturity, come in mash or pellet form. They may include either loose or pelleted mash plus whole or cracked grain, sometimes referred to as scratch grains. A commercial concentrate may also be combined with ground grain or with soybean meal and ground corn in the proportions recommended by the manufacturer. Once again, the small grower will find it advantageous to use a complete ready-mixed mash when the birds are reared in confinement. If they are on range, a complete feed — preferably in pellet form, supplemented with good range, grains, and insoluble grit — makes a sound feeding program. Growth-rate and feed-consumption information for heavy and light roaster turkeys and water-consumption information are given in the following charts.

Growth Rate and Feed Consumption of Rapid-Growing Roaster Turkeys

AGE (WEEKS)	LIVE WEIGHT (POUNDS)		FEED CONVERSION (POUNDS FEED/POUNDS GAIN)	
	TOMS	HENS	TOMS	HENS
2	0.78	0.67	1.22	1.28
4	2.63	2.22	1.38	1.45
6	5.45	4.46	1.53	1.64
8	9.46	7.57	1.65	1.78
10	13.97	10.95	1.77	1.94
12	18.90	14.27	1.91	2.11
14	24.07	17.23	2.07	2.28
16	29.14	20.02	2.23	2.45
18	34.14	22.21	2.41	2.66
20	38.84	25.00	2.59	2.87
22	43.27		2.78	

Source: Nicholas Turkeys Technical Information. Average of two strains of turkeys. Sonoma, CA: Nicholas Turkey Breeding Farms.

Estimated Daily Water Consumption of Roaster Turkeys (Gallons per 100 Birds)

AGE IN WEEKS	WATER CONSUMPTION
1	1
2	2
3	3
4	4
5	5
6	6
7	7.5
8	9.5
9	11.0
10	12.5
11	14.0
12	15.0
13	16.0
14	16.5*
15	17.0*
16	16.5*
17	16.5*
18	16.5*
19	16.5*
20	16.5*

*Consumption will vary from 15 weeks to maturity from 14 to 19 gallons (52.9–71.9 L) per 100 birds per day, depending upon environmental temperatures.

Source: Dr. Salsbury Laboratories, Charles City, IA.

Range Rearing

It is possible to reduce the cost of rearing turkeys by putting them on range. This is especially true if the diet can be supplemented with home-grown grains. Turkeys are good foragers, and if good green feed is available on the range, this will help supplement the diet and reduce the cost of the feeding program.

Disadvantages

Range rearing is not without its problems. Losses are possible from soil-borne diseases, insects, thieves, diseases from migratory and other birds, adverse weather conditions, and predators. Because of these factors, confinement rearing, which tends to minimize these losses, has rapidly replaced range rearing in most areas, and from a biosecurity standpoint, is recommended.

Portable Shelters

Portable range shelters that can be moved to various areas of the range are preferred with layer flocks. The shelter should provide about 1½ square feet (0.1 sq m) of floor space per bird. It should be equipped with roosts, or slat floors with the slats located 1½ inches (3.8 cm) apart.

Poult Age and Range Selection

Normally, you can put turkey poults out on the range at 4 to 8 weeks of age. The flock should be well feathered, especially over the hips and back, before being put out. Check the weather forecast for several days before putting the birds on range. It is best to move them in the morning, as this helps get them used to range conditions without losses.

A range area that has been free from turkeys for at least 1 year, and preferably for 2 years, is desirable. Poorly drained soil should be avoided, since stagnant surface water may be a factor in turkey disease outbreaks. Use a temporary fence to confine the flock to a small area of the range, and move it once a week, or as often as the range and weather conditions indicate. Provide artificial shade if there is no natural shade. Several rows of corn planted along the sunny side of the range area provide good shade as well as some feed as the corn matures. About 1 acre (0.4 ha) of good pasture is required for 250 turkeys. If range shelters are used, it is best to move them every 7 to 14 days, depending upon the weather and the quality of range. Also move feed and watering equipment as needed to avoid muddy and bare spots.

It is possible to use a combination of confinement and range rearing by providing a permanent shelter, such as a barn, with fenced-in range areas around it. These fenced-in areas can be used on an alternative basis every year or two.

Range Crops

Range crops will depend upon the climate, soil, and range management. Many turkey ranges are permanently seeded; others are part of a crop-rotation plan. As part of a 3- or 4-year crop rotation, legume or grass pasture and an annual range crop such as soybeans, rape, kale, sunflowers, reed canary grass, or Sudan grass have been used successfully. Sunflowers, reed canary grass, and Sudan grass provide green feed and shade. For permanent range, alfalfa, ladino clover, bluegrass, bromegrass, and others are satisfactory.

Range Feeders, Waterers, and Shelters

If you provide feeders outside, they should be waterproof and windproof, so that the feed is not spoiled or blown away. They should be placed on skids, or be small enough to move by hand. Trough feeders are inexpensive and relatively easy to construct. You can also purchase specialized turkey feeding equipment. To prevent excessive feed waste, all feeding equipment should be designed so that it can be adjusted as the birds grow. The lip of the feed trough should be approximately on line with the bird's back and the feed level in the trough kept at about half to prevent waste. Pelleted feeds are less likely to be wasted on range.

Provide two 10-foot (3 m) feeder troughs per one hundred birds if the flock is hand-fed each day. When bulk feeders are used, feeder space should conform to the equipment manufacturer's recommendations. Provide one 4-foot (1.2 m) automatic trough waterer or equivalent per one hundred birds. The waterers should be cleaned daily and disinfected weekly, and, if possible, they should be outfitted with portable shade or situated to take best advantage of natural shade.

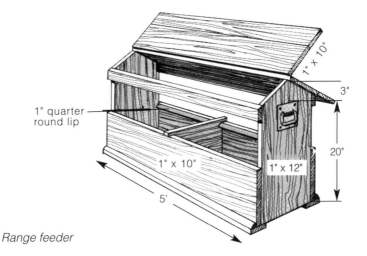

Range feeder

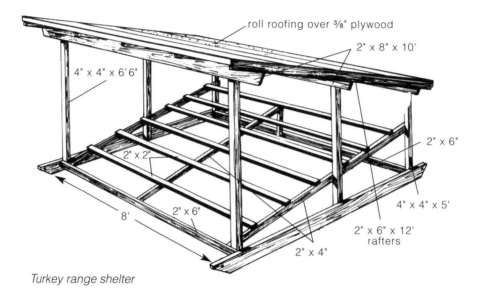

Turkey range shelter

Ensure that range shelters provide roosting space and cover for the birds during adverse weather and hot sun. During the hot period of the year, particularly when birds are close to marketing time, it is advisable to pull extra shelters onto the range to provide more shade for the birds when natural shade is not available.

Grooming

The following grooming procedures are often performed to prevent injury to birds and thus improve their condition when they go to market.

Beak Trimming

It is recommended that the poults be beak trimmed when they are from 2 to 5 weeks of age. Delaying beyond this period makes it difficult to handle the heavier birds and excessive feather picking may occur. To prevent cannibalism and feather pulling, beak trimming should be a regular practice before noticeable picking occurs.

The poults should not be beak trimmed at day of age at the hatchery, as this can interfere with their ability to eat and drink and may cause

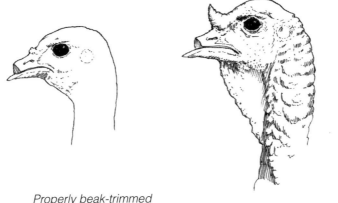

*Properly beak-trimmed
young turkey*

Properly beak-trimmed adult turkey

starve-outs and dehydrated birds. The beak-trimming job is done with an electric beak trimmer and about half of the upper beak should be removed, being careful not to cut into the nostril. Make sure that feed and water levels are deep enough for the birds. The illustrations show the correct beak-trimmed appearance of turkeys.

Desnooding

The snood (a fleshy appendage on the top of the head/back of the beak) is sometimes removed. Removal of the snood helps prevent head injuries as a result of fighting or picking, and may prevent erysipelas infection from getting started in the flock. You can remove the snood at 1 day old by pinching between the finger and the thumbnail. You can also cut it with fingernail clippers or scissors up to approximately 3 weeks of age.

Wing Clipping or Notching

The feathers of one wing are sometimes cut off with a sharp knife to prevent the birds from flying, if on range. Wing notching, or the removal of the end segment of one wing with a beak trimmer, is another method sometimes used to prevent flying on range. You can notch the wing with an electric beak trimmer from 1 day of age to 10 days. Wing clipping and wing notching are not used as much as they once were because of injuries resulting from the birds' attempts to fly. This may cause carcass bruises and detract from the dressed appearance.

Toe Clipping

Toe clipping is frequently done to prevent scratches and tears of the skin on the birds' backs and hips. It is especially helpful when birds are crowded or nervous but appears to help even when they are on the range. The two inside toes are clipped so that the nails are removed. Surgical scissors or an electric beak trimmer may be used. This should be done at the hatchery.

General Management Recommendations

Isolate young poults from older turkeys. Take care not to track disease organisms from the older stock to the younger. Follow a good control program for mice and rats. These rodents not only are disease carriers but also consume large quantities of feed. Rats can kill young poults.

Ideally, no other avian species, such as chickens, game birds, or waterfowl, should be on the same farm. However, with the medications available today, it is possible to grow small flocks of turkeys with other types of birds, if they are kept separate. It is still somewhat risky. If abnormal losses or disease symptoms occur, promptly take the birds to a disease diagnostic laboratory. One of the first symptoms of a disease problem is a reduction in feed and water consumption. Without good records on daily feed consumption, it is difficult to detect changes in intake. Dead birds should be disposed of immediately in a sanitary manner, by incinerating or composting, or by use of a disposal pit if allowed in your area (see chapter 13).

With good management, you should be able to raise to maturity 95 percent of the turkeys started. With high poult costs and feed costs, mortality can become expensive, especially when birds are lost in the latter part of the growing period.

Disease Control

Depending upon the disease exposure in your area, it may be necessary to vaccinate the poults for such diseases as Newcastle, fowl pox, erysipelas, and fowl cholera. In planning a vaccination program for your flock, check with the poultry diagnosticians at your state laboratory, your county Extension agent, or some other knowledgeable person. Small flocks often are not vaccinated.

Medication is effective in reducing losses from such diseases as coccidiosis and blackhead when used in the feed at preventive levels. Antibiotics and other drugs are of value in preventing and treating diseases, but such medications should not be used as a substitute for good management.

The Production of Turkey Hatching Eggs

Occasionally, you may want to mate a few turkeys for the production of hatching eggs. The turkey breeding flock not only is somewhat more difficult to manage than the chicken breeding flock, but it also is considerably more expensive. The cost of growing the bird to maturity is several dollars (see page 160), and the amount of feed required to maintain the hens and toms during the holding and breeding periods is also substantial.

Approximate Feed Consumption of Turkey Breeders (Pounds per Bird per Day)

TYPE OF TURKEY	HENS	TOMS
Large	0.60	1.00
Medium	0.45	0.75
Small	0.35	0.65

Source: Turkey Production AGR Handbook *No. 393, USDA.*

Selecting the Breeders

Start the turkeys to be kept as breeders about 8 months before egg production is desired. Those that come into production at too early an age lay small eggs. Fertility tends to be poorer in small eggs.

Select only the best birds as breeders. Look for candidates with good, full breasts (those with no protruding keel bones). Select for strong, straight legs, straight keel bones, and straight backs. Use only those birds that are healthy and vigorous for breeders.

Mating

One tom per ten hens is recommended for the older varieties of the light breeders. Most of the heavy commercial-strain breeders do not mate naturally, and artificial insemination is required.

Lighting

The usual procedure is to commence lights on the toms about 4 to 5 weeks prior to mating. Toms need this light period to stimulate them to produce sperm. Hens should receive light about 3 weeks prior to the onset of egg production.

A 13- to 15-hour light period, artificial and natural combined, should be used. A 50-watt incandescent bulb or a 9-watt compact-fluorescent (CF) lamp for each 100 square feet (9.3 sq m) of floor space is adequate. A timer can be used to bracket daylight hours with artificial light.

Egg Production

Turkeys are not as good layers as most chickens. Modern commercial heavy strains may be expected to produce eighty to one hundred eggs during a 20-week production cycle. Medium-sized turkeys and light birds may lay anywhere from seventy-five to more than one hundred in the same period.

First-year egg production is best. After that, production diminishes at the rate of approximately 20 percent each succeeding year. Egg size increases with age; hatchability tends to decrease.

Mating Habits

Females will not mate until they commence egg production. Toms must be ready for semen production when the hen is ready to mate. The toms strut most during the mating season. When the hen is ready to mate, it approaches the tom of its choice. The hen squats near the tom, the male mounts it, and copulation usually takes place. Some toms may never mount the hen; others may mount but not complete the insemination. When the male performance is not satisfactory, the hen may become disinterested and not mate for a period of time. If this happens, low fertility can result.

Fertilization

Fertilization of the egg takes place in the upper part of the oviduct. Sperm is stored there for at least several days and can fertilize eggs for as long as 3 weeks after insemination. Sperm is also stored in sperm nests at the uterovaginal junction. Storage life of the sperm shortens as the hens become older.

Artificial Insemination

Modern strains of commercial breeder hens are fertilized totally by artificial insemination. The semen used for artificial insemination is obtained by milking the toms. Semen can be collected from the toms two or three times per week. Stimulate the tom by stroking its abdomen and pushing the tail upward and toward the head. The male copulatory organ enlarges and partially protrudes from the vent. Grip the rear of the copulatory organ with the thumb and forefinger from above, fully exposing the organ, and squeeze the semen out with a short, sliding, downward movement. The males soon become trained and ejaculate readily when stimulated. Collect the semen in a small glass beaker or a test tube. The tom yields 0.2 to 0.5 milliliter. The semen should be creamy and free from fecal material. Some producers withhold feed and water from toms for 8 to 10 hours to help avoid fecal contamination of the semen.

Semen should be used within 30 minutes. About 0.02 milliliter of semen per hen is adequate. Usually, good fertility will result when hens are inseminated two times at 4-day intervals at the onset of egg production. Insemination every 2 or 3 weeks thereafter is usually desirable unless fertility is high in the mated flock.

The hen is inseminated by exposing the opening of the oviduct and inserting a small syringe (usually a 1 mL syringe) without a needle into the oviduct about 1½ inches (3.8 cm). The oviduct is exposed by pushing outward and exerting pressure on the abdomen, while at the same time forcing the tail upward toward the head. The oviduct can be exposed or protruded only in those hens that are in laying condition.

Broodiness

Broodiness is an inherited characteristic. Turkeys are inclined to be more broody than most breeds of chickens, and some strains of turkeys are more broody than others. It is difficult to identify broody turkeys. If you find the birds on the nest in the early morning or evening, it is safe to assume they are broody. Broody birds should be driven into a broody pen or yard and left there for about 5 days. They should be fed and watered in the broody pen. Dark areas and areas where they can nest should be eliminated. Drive broody birds out of the nests, and close the nests at night. Slat or wire floors are best for broody pens. One or more toms in the pen will tend to keep the hens on their feet and discourage broodiness. Hens usually want to mate a few days after they go broody, so this practice may also improve fertility.

You can discourage broodiness by moving the birds to different areas, providing roosts, and chasing birds off the nest when gathering eggs. Gathering eggs several times each day may also help prevent broodiness.

Housing

Buildings used for brooding young stock are satisfactory for breeders. Confinement is preferred to range management, where unfavorable weather is common or where predators may be a problem. Houses should be well insulated and ventilated for maximum comfort in cold or warm weather. The floor can be concrete, dirt, asphalt, or wood. It should be covered with a good litter. Houses should be thoroughly cleaned and disinfected between flocks.

More floor space is required for breeders: 6 to 8 square feet (0.55–0.74 sq m) per bird is usually recommended. Where only females are housed, 5 to 6 square feet (0.46–0.55 sq m) is adequate for large turkeys and 4 to 5 square feet (0.37–0.46 sq m) for small ones.

In warm climates or where winters are mild, you may give breeders access to fenced range or yard. They should have a shelter to protect them from bad weather and predators. Keep the range area well drained and provide about 150 square feet (13.9 sq m) per bird. If a yard is used, allow 4 to 5 square feet (0.37–0.46 sq m) of area per bird. Equip the shelter with roosts and provide at least 4 square feet (0.37 sq m) of floor space per bird. The feeders, waterers, and broody pen are best located inside the shelter.

Nests

Provide one nest for every five birds. Nests should be located in a dark area of the house and be available prior to the time egg production starts, so the hens will get used to them.

Some breeders prefer tie-up or trap nests, but open nests may be used. Nests should be 2 feet by 2 feet (61 cm × 61 cm) if they are the manger or open type. If nests are placed outside the shelter, they should be protected by a roof.

Feeding

Put breeders on a special turkey breeder diet 1 month before egg production is expected. Feed according to the manufacturer's instructions. All-mash diets are preferred, but mash-and-grain diets are satisfactory if fed in the right proportions. If too much grain is fed in relation to the mash, egg production and hatchability may be reduced. If used properly, this method has the advantage of making use of homegrown grain without having to grind or mix it. Homegrown grains can be used in an all-mash program by grinding and mixing with a concentrate at home or having them mixed elsewhere.

Mash or concentrate is available in a pelleted form. There is less waste with pelleted feeds, particularly if fed on range. An insoluble grit or gravel should be available for confinement breeders. When the birds are fed and watered outdoors, move the equipment frequently to avoid muddy spots and possible sources of disease.

Six linear inches (15.2 cm) of feeder space should be provided for large turkeys and 4¼ linear inches (11.4 cm) for the small birds. Provide two automatic waterers per one hundred birds. One linear inch (2.5 cm) of water space per hen is adequate. Turkeys normally consume 2 pounds (0.9 kg) of water for every pound (0.4 g) of feed. Lack of water can be more harmful to the birds than the lack of feed.

Care of Eggs

Gather the eggs at least three times per day, more frequently if the birds tend to crowd certain nests. Frequent gathering will help avoid breakage, dirty eggs, and possibly frozen eggs or exposure to high temperatures.

If eggs are cleaned, they must be cleaned properly, using a detergent sanitizer made for egg washing. A washing temperature of 110 to 115°F (43–46°C) is recommended. Excessively dirty eggs gathered in hot, wet weather can easily become contaminated. Contaminated eggs often explode in the incubator.

If eggs are to be held before incubating, they should be turned daily. When eggs are turned and held under good storage conditions, hatchability may be good for 2 weeks. Place eggs in filler flats or cartons and elevate the container at one end to slant the eggs about 30 degrees. At least once each day, shift ends with the container and elevate the opposite end. You will improve hatchability by doing this.

Flock Health

Hens should be beak trimmed when placed in the breeding pens. Remove about half of the upper beak (see page 176). Maintain feed and water levels at a height that will permit the beak-trimmed birds to eat and drink.

Some states require that breeders be blood-tested for pullorum disease and fowl typhoid, and possibly other diseases such as chronic respiratory disease, paratyphoid, and infectious synovitis. Follow the recommendations of your county agent, state Extension Poultry Specialist, or diagnostic laboratory.

The key to good flock health is prevention. Keep feeders and waterers clean, keep pens clean and dry, and use biosecurity measures to keep from spreading disease from one flock to another.

TOM'S TURKEYS

Cathy and her husband Tom always wanted to raise some turkeys so they could call them "Tom's Turkeys" and have their own fresh turkey for Thanksgiving. They picked up one of the "How to Raise a Small Flock of Turkeys" pamphlets from the feed store and built a turkey pen and shelter. They purchased five turkeys and starter feed from the feed store in mid-June and grew the birds. The salesman at the feed store told them it was easy to process the turkeys, that people did it all the time, so they never gave that part of the cycle another thought.

They were attentive to their flock and the birds grew fine and healthy through the summer and early fall. Then came the day of reckoning. It was the Tuesday afternoon before Thanksgiving. They hung a rope from a tree branch to hang the birds upside down for slaughter, set up a metal garbage can with hot water to scald the birds and a table to eviscerate them, and filled a big plastic garbage can with a plastic-bag liner with cold water for chilling, just as the pamphlet said. They sharpened a couple of knives and were ready to go. After the first bird, they were almost ready to give up. Pulling feathers by hand was no picnic, and evisceration did not look as bad in the black-and-white pamphlet as it did in living color.

Not willing to give up, they called the Extension specialist at the local university. He talked them through it and they finished their five birds. In talking with the poultry specialist, they found out that there were several other families in their area who grew just a few turkeys each year and who also wished there were a place nearby to have the birds processed. On the advice of the poultry specialist, the families got together, formed a small co-op, and purchased a scalding tank, feather picker, killing cones, and stunning equipment. They also purchased bulk feed at a much better price than bag feed and split it among the families. The Extension specialist held a processing clinic for them about a month prior to the next Thanksgiving and taught them how to work as a team processing their turkeys.

Now Tom's Turkeys raises thirty-five birds a year, and among all the members of the co-op, they raise and process 250 turkeys each year, selling fresh birds in the local area.

TEN

WATERFOWL PRODUCTION

There are almost as many reasons for keeping ducks and geese as there are breeds to select from. For one thing, waterfowl can be raised with relative ease. They are hardy and not subject to as many diseases as are other types of poultry. Ducks will eat slugs and snails, while geese will control your overgrown grass and weeds. Geese also can be used for guard duty around your home or farm. Numerous small flocks of waterfowl are grown to grace a small pond, as a hobby, or for exhibition. Not to be overlooked is the fact that properly finished and dressed, waterfowl are excellent for food. A number of people, in fact, prefer waterfowl to other types of poultry for the Thanksgiving and Christmas holidays. In addition, the eggs from waterfowl may be used for decorating or eating and are especially good for baking. Some breeds of ducks are excellent egg producers, outproducing their chicken cousins.

Ducks

Most of the purebred ducks being raised in North America today are one of sixteen breeds. Most of these breeds are descended from Mallards and are distinguished by their color or plumage patterns. The breed you select will depend upon the purpose for which the birds are to be raised — meat, eggs, general purpose, or bantams. People who keep ducks for ornamental purposes, for a hobby, or for exhibition may choose the Cayuga, Crested, Swedish, or Buff ducks. Some choose the smaller duck varieties, or bantam ducks. Included in this group are the Calls, East Indie, Wood Duck, and Mallard.

Meat Breeds

Meat breeds are distinguished by their large body size, fast growth rates, and calm disposition. Some of the best choices include White Pekin, Aylesbury, Rouen, and Muscovy ducks.

White Pekin

The White Pekin is probably the most common breed of domesticated duck in North America. It was introduced into the United States from China in the 1870s and is well suited for the production of meat. The birds produce good-quality meat and reach a market weight of about 7 pounds (3.2 kg) in less than 8 weeks when properly managed. They are a large, white-feathered bird. The bill is an orange-yellow color; their legs and feet, a reddish yellow; and they have yellow skin. The adult drake weighs 9 pounds (4.1 kg); the adult duck, 8 pounds (3.6 kg).

White Pekin

The White Pekin is a fairly good egg producer. The average yearly production reaches approximately 160 eggs. However, they are generally poor setters and seldom raise a brood of ducklings.

One drawback to the Pekin is the "talkative" nature of the female, which may annoy neighbors. Some people also object to the fattiness of the carcass.

Aylesbury

Aylesbury

The Aylesbury originated in England and, like the White Pekin duck, is a good meat bird. It reaches market weight in about 8 weeks. The Aylesbury has white feathers; pinkish white skin; a pale flesh-color, long, straight bill; and bright light orange legs and feet. The eggs are white or off-white. The adult drake weighs 9 pounds (4.1 kg); the adult duck, 8 pounds (3.6 kg). It lays somewhat fewer eggs than the White Pekin and is a poor setter.

Aylesburys are rather slow moving but very tame, and are good fat roasters when well fed on concentrate finisher feed. For maximum fertility and hatchability, breeders must have swimming water and be fed a ration that contains 18 to 20 percent protein and is fortified with vitamin-rich ingredients such as alfalfa meal and cod-liver oil or a commercial vitamin premix.

Muscovy

Muscovy

These birds are native to South America, and there are several varieties of Muscovies, the white being the most desirable for market purposes. They produce meat of excellent quality and taste when marketed before 17 weeks of age. They are relatively poor egg producers but good setters. The adult drakes weigh 10 pounds (4.5 kg); the adult ducks, 7 pounds (3.2 kg). Muscovies have white skin.

Muscovies have several distinguishing features. Their faces are covered with caruncles (rough, red skin), and drakes have caruncles around the base of their bills. Both sexes have narrow head crests that can be elevated and lowered at will. Hens are almost mute and the voices of males sound like a muffled hiss. The drakes can be almost twice as heavy as the females and have larger facial-skin patches, making it easy to distinguish the gender of mature birds.

Muscovies are great foragers and can utilize larger quantities of grass than other ducks.

Mules and Hinnies

Muscovies can be mated with other breeds of ducks to produce some interesting offspring. If you cross a Muscovy male with a Pekin female, you will hatch mules; if you cross a Pekin male with a Muscovy female, the offspring are called hinnies.

In Europe and the East mule ducks are produced because of their large size, good-quality liver, and reduced carcass fat. Because natural mating between these breeds results in poor fertility, only about 20 to 30 percent, artificial insemination (AI) is frequently used with ducks. With AI, the fertility reaches almost 80 percent. Another interesting point is that whereas a Pekin takes 28 days to hatch and a Muscovy takes 35 days, a mule or a hinny takes 32.

Approximately 60 percent of mule ducks are males. Some of their characteristics are similar to those of the Muscovy, since they are large, quiet, slow moving, and have long claws; but they are also like Pekins because they swim well, the males and the females are about the same size, and they do not fly. Although hinnies are not grown anywhere commercially, they are interesting. Male hinnies are much larger than females, like the Muscovy, yet the females look like Pekins and fly quite proficiently.

Mules and hinnies cannot reproduce, because males of each cross are sterile and only the hinny females lay eggs (though they cannot hatch). However, if you have Muscovy and Pekin together, the chances are poor that they will mate; but if they do, a hinny will probably be the result, since Pekin males can catch Muscovy females more easily than Muscovy males can catch Pekin females.

Rouen

The Rouen duck, originally from France, reminds one of the Mallard. It has the same striking color patterns but is typically several shades darker than the Mallard, and larger. The adult drake weighs 10 pounds (4.5 kg); the adult duck, 9 pounds (4.1 kg). The Rouen's pigmented plumage gives it a less desirable dressed appearance. It is excellent for

Rouen

home consumption, where the dressed appearance is not so important.

Egg-Producing Breeds

If your interest is in a breed of ducks to produce eggs, your choice should be the Bali, Khaki Campbell, or Indian Runners.

Bali

Although a most unusual breed, it is one of the oldest domesticated ducks, originating in Bali and other Indonesian islands. Some believe that the Indian Runner was developed from the Bali.

Bali

The Bali has an erect carriage with a long body that is somewhat heavier than that of the Indian Runner. On the top of its head is a medium-sized globular crest. The Bali will lay between 150 and 250 eggs per year, depending upon management. The Bali are rare and not often found outside their homeland.

Khaki Campbell

The Khaki Campbell was developed in England. Several varieties have been selected for high egg production. Some are reported to average close to 365 eggs per duck in a laying year. It is interesting to note that this is a higher rate of production than those of the highest-producing strains of chickens.

Khaki Campbell

The Khaki Campbell male has a brownish bronze lower back, tail coverts, head, and neck, and the rest of its plumage is khaki. The beak is greenish black and the legs and toes are brown. The adults weigh only 4½ pounds (2 kg); thus, the Khaki Campbell is not known for production of meat.

Indian Runner

Although the Indian Runner originated in the East Indies, its egg-production qualities were developed in Europe. Indian Runners are second only to Khaki Campbells in egg production. There are three Indian Runner varieties, White Penciled, Fawn, and White. All of them have orange to reddish orange feet and shanks. The males and its females weigh approximately 4½ pounds (2 kg). They are not good meat birds.

Indian Runner

Breeds of Geese

The most popular breeds for meat production are the Toulouse, Embden, and African geese. Other common breeds in the United States include the Chinese, Canada, Pilgrim, Buff, Sebastopol, and Egyptian.

In choosing a breed, consider the purpose for which the geese are to be raised. You may raise geese for either meat or egg production, or as weeders, show birds, or farm pets. Some of the crossbreeds, such as the cross between the White Chinese male and the medium-sized Embden female, usually result in fast-growing white geese of good market size.

Toulouse

The Toulouse has a broad, deep body and is loose feathered, a characteristic that helps give it its large appearance. The plumage is dark gray on the back, gradually shading to a light gray edge with white on the breast and white on the abdomen. The bill is pale orange and the shanks and toes are a deep reddish orange.

Toulouse

Embden

The Embden is a pure-white goose. It is a much tighter-feathered bird than the Toulouse and, therefore, appears more erect. Egg production averages from thirty-five to forty eggs per bird. The Embden is a better setter than the Toulouse and is one of the most popular breeds for meat. It grows rapidly and matures early.

Embden

African

The African has a distinctive knob on its head. The head is a light brown, the knob and bill are black, and the eyes dark brown. The plumage is ash brown on the wings and back and a light ash brown on the neck, breast, and underside of the body. It is a good layer, grows rapidly, and matures early. It is not as popular for meat production because of its dark beak and pinfeathers.

African

Chinese

The Chinese are smaller than the other standard breeds. There are two varieties, the Brown and the White. Both mature early and are better layers than the other breeds. They average from forty to sixty-five eggs per bird annually. The Chinese grows rapidly. It is an attractive breed and makes a desired medium-sized meat bird. It is popular for exhibition and ornamental use.

Chinese

Canada Goose

The Canada Goose is the common wild goose of North America. Its weight ranges from about 3 pounds (1.4 kg) for the lesser Canada Goose to about 12 pounds (5.4 kg) for the giant or greater Canada Goose. The Canada is a species different from other breeds of geese. It can be kept in captivity only by close confinement, by wing clipping, or by pinioning the wings. They may be kept only by permit, which must be obtained from the Fish and Wildlife Service of the U.S. Department of the Interior in Washington, D.C.

Canada Goose

The breed does not have the economic value of other breeds of geese — they mate only in pairs, are late maturing, and lay few eggs.

Pilgrim

The Pilgrim is a medium-sized goose and is good for meat production. The males and females of this breed can be distinguished by feather color. In day-old goslings, the male is a creamy white and the female is gray. The adult male remains all white and has blue eyes. The adult female is gray and white and has dark hazel eyes.

Pilgrim

Buff

The Buff has only fair economic qualities as a market goose and only a limited number are raised for meat. Color varies from dark buff on the back to light buff on the breast, and from light buff to almost white on the underpart of the body.

Buff

Sebastopol

The Sebastopol is a white ornamental goose that is attractive because of its soft plumelike feathering. The breed has long curved feathers on its back and sides and short curled feathers on the lower part of its body.

Sebastopol

Egyptian

This long-legged but very small goose is kept primarily for ornamental or exhibition purposes. Its coloring is mostly gray and black with touches of white, reddish brown, and buff.

Egyptian

Getting Started

If you wish to raise a small flock of ducks or geese, you can best get started by purchasing day-old ducklings or goslings. This eliminates the need for keeping breeding birds and incubating the eggs. Day-old stock is available in most areas of the country. A list of the hatcheries selling waterfowl is printed in the National Poultry Improvement Plan Hatchery List. This can be obtained by writing to the U.S. Department of Agriculture in Washington, D.C. Other good sources of information concerning the availability of young birds include the state Extension Poultry Specialist and the county Extension agent in your area.

Brooding and Rearing

As is the case with other types of poultry, waterfowl must be put on feed and water as soon as they are placed under the brooder. Young ducklings, and particularly young goslings, are somewhat hardier than chickens and turkeys. A special brooder building is not necessary for brooding small numbers of geese. However, they do need to be kept where it is warm, dry, free from drafts and reasonably well lighted. The building should be designed so that it can be ventilated without chilling the birds. The building also must be protected against dogs, cats, rodents, and other pests and predators. Almost any type of building that will provide these conditions is suitable for brooding ducklings or goslings.

Facilities

Clean and disinfect the brooder house well before the birds arrive. When they arrive, have the house, the litter, the feeding and watering equipment, and the brooder stove ready. The brooder should be operating at least 1 day before the birds arrive. The first day, place a brooder guard approximately 2 feet (0.6 m) from the hover. Move it back gradually and remove it at about 7 days. The brooder guard will confine the birds to the brooding area and prevent huddling and chilling before they are able to adjust to the location of the heat source.

You may start waterfowl on wire, slat, or litter floors. The most practical method is to start them on litter. Wood shavings, sawdust, chopped straw, and peat moss are all good litter materials. The material should be free from mold, since moldy litter can cause mortality. The proper equipment arrangement for brooding is shown on page 82.

The birds should have adequate floor space. Ducklings need ½ square foot (464.5 sq cm) of floor space per bird the first week, ¾ square foot (696.8 sq cm) the second, and 1 square foot (929 sq cm) the third week. If they are housed in confinement, you will need to increase this space allotment to 2½ square feet (0.2 sq m) by the time they reach 7 weeks of age.

Goslings will require more floor space. They should have ½ to ¾ square foot (464.5–696.8 sq cm) of floor space the first week and 1 to 1½ square feet (0.09–0.14 sq m) the second. Increase the amount slightly until they go on range. You can place them on range at 2 to 4 weeks of age. No shelter is needed, but some type of shade should be provided.

Heating

Ducklings normally need supplementary heat for approximately 4 weeks. During warm weather or in warm climates, heat may be needed for just the first 2 or 3 weeks. Goslings usually can get by without heat after 2 weeks of age. Waterfowl feather rapidly, and thus do not require as long a brooding period as do baby chicks. Any brooder unit suitable for brooding chicks or turkeys is satisfactory for ducklings or goslings. Electric, gas, oil, coal, and wood-burning brooder units are all suitable.

Infrared Lamps

Infrared heat lamps are excellent for brooding small groups of birds. One infrared lamp for as many as thirty birds is satisfactory. To determine the number of ducklings or goslings a hover brooder will accommodate, cut the brooder's rated chick capacity in half. They require 13 to 14 square inches (83.9–90.3 sq cm) of hover space per bird.

Temperature

Keep the brooder temperature in the vicinity of 85 to 90°F (29–32°C) the first week, and reduce it approximately 5°F (3°C) per week during the next few weeks or until 70°F (21°C) is reached. The behavior of the young birds is the best guide to the temperature required (see page 83). When the temperature is too hot, the birds will crowd away from the heat. High temperatures may result in slower weight gain and slower feathering. When the temperature is uncomfortably cold, the birds will tend to huddle together under the brooder or, perhaps, crowd into corners. When a hover brooder is used, a night-light will tend to discourage crowding; when infrared brooders are used, enough light is provided to make extra light unnecessary. When the temperature is just right, the birds will be well distributed over the floor and using all the feeders and waterers.

In warm weather, goslings can go outdoors as early as 2 weeks of age, but will need frequent attention until they learn to go back into the coop or brooder when it rains. Goslings must be kept dry to prevent chilling, which can result in piling and smothering. Housing is usually not needed after the geese are 6 to 8 weeks old.

Feather Pulling

As with other types of poultry, feather pulling can be a problem with ducks. If this vice starts, give the birds additional space. It may be necessary to trim the bill of the birds if the problem continues. This is done by nipping off the forward edge of the upper bill with an electric beak-trimming machine.

Feeding and Watering Space

Waterfowl will consume more water per pound of body weight than chickens or turkeys, and thus it is important to keep clean drinking water in front of the birds at all times. You may use hand-filled water fountains or automatic waterers. Place the water fountains on wire or slatted platforms or over a screened drain. This will help keep the litter dry. Use only waterers that the birds cannot get into. They should be cleaned daily.

Give waterfowl access to range or a yard when they are old enough to tolerate the weather conditions. Ducklings will manage nicely on range at 4 weeks of age unless the weather is cold. You can place goslings outdoors at 2 weeks of age, weather permitting. They need shade and cannot tolerate chilling rains until they are well feathered on the back. When birds reach 5 to 8 weeks of age, they need shelter only during extreme weather conditions. Most commercial growers provide ponds for their birds at approximately 5 weeks of age; however, waterfowl can be raised without swimming water.

Feeders

There are several good types of feeders for waterfowl. Feed troughs, such as those used for chickens (see pages 25–27) work well. You can use small feeders until the birds are 2 weeks of age; large feeders should be used for older birds and breeding stock. Cover feed hoppers that are used outdoors to protect the feed from wind and rain. A range feeder designed for chickens or turkeys will work well. A plan for a range feeder is also shown in chapter 2 (see page 27).

To start ducklings and goslings off, it is best to use chick-box covers, shallow pans, or small chick feeders. Hanging tube feeders are excellent. One of these with a pan circumference of 50 inches (1.3 m) is suitable for fifty ducklings or twenty-five goslings for the first 2 weeks.

Waterers

Pans or troughs are satisfactory waterers. Equip them with wire guards or grills to prevent the birds from playing in them and spilling water. Place the waterers over low wire-covered frames, preferably over drains to prevent the litter from getting too wet. (A satisfactory water stand for waterfowl is shown on page 30.) Increase the size and number of waterers as the birds become older. Approximately 4 feet (1.2 m) of trough is good for 250 young waterfowl for the first few weeks. The waterer should be wide enough to permit the bird to dip its bill and head into it.

Feeding

Some commercial feed companies formulate rations specifically for ducks and geese. In a number of areas, however, specially prepared rations will not be available. It is possible to mix your own feed, but before doing so, write to your Extension Poultry Specialist for formulas. If feeds for ducks are not available, they may be started on crumbled or pelleted chick starter for the first 2 weeks. Pellets are recommended, because they are easier to consume. They also reduce waste and do not blow around like mash when used outdoors. Feed conversion is usually better with pellets; however, the lack of pelleted feed should not discourage a grower who wishes to produce ducks or geese on a small scale. Satisfactory results can be obtained with mash. If you use a chick starter, it should not contain any drugs that may be harmful to the ducklings. The use of a medicated feed is not usually necessary. Coccidiosis is not nearly the problem in ducklings that it is in chickens, though it does affect some flocks.

Regardless of whether you purchase feed or mix it on the farm, store it away from rodents and insects in a clean, dry place to prevent contamination and mold growth. A plastic can inside a covered metal can makes an excellent storage place for feeds. Never use moldy feeds, because they produce toxins and ducks are extremely sensitive to mold toxins. Use feed within 3 weeks of manufacture in hot, humid weather to prevent loss of vitamin activity and mold growth.

Feeding Ducks

Ducklings should be fed a starter diet in the form of ⅛-inch-diameter (0.3 cm) pellets or crumbles from hatch to 2 weeks of age. After 2 weeks, a grower diet of ³⁄₁₆-inch-diameter (0.5 cm) pellets is recommended.

Ducklings do well on a starter diet that contains 20 percent protein. From 3 to 9 weeks, they can be fed a 16 to 18 percent protein diet. Birds more than 9 weeks of age can be fed 13 to 16 percent protein if grown for meat or 16 percent protein if to be used as breeders. If a special grower diet is not available, a chick grower is satisfactory. When ducks are to be dressed as green ducks — that is, at 7 to 9 weeks of age — this diet can be fed for the full period. In addition to a pelleted grower ration, cracked corn or other grains are often included in the diet for goslings. Keep feed before the birds at all times, and provide an insoluble grit. Ducks are not as good foragers as geese but may be put on pasture at 4 weeks of age if they are to be kept beyond the 7- to 9-week period. Feed-consumption and live-weight information for ducks and geese are shown in the following charts.

For the backyard or small duck flock without access to commercially produced waterfowl feeds, chicken feeds will work. A 22 to 23 percent protein chick starter can be used for the first 2 weeks, followed by a 16 to 18 percent grower diet. An 18 percent broiler finisher may also be fed to fatten up a duck grown for meat.

For pet ducks kept year-round, the addition of cracked corn, as much as 10 to 15 percent of their total intake, will provide extra energy for the colder winter months.

Feeding Geese

Goslings can be fed a 20 to 22 percent waterfowl feed or goose starter in the form of ³⁄₃₂- or ³⁄₁₆-inch (2.4–4.8 mm) pellets for the first 6 weeks. After 6 weeks, feed a 15 percent goose grower in the form of ³⁄₁₆-inch (4.8 mm) pellets. However, geese can be fed pellets, mash, or whole grains. The main objective is to meet the nutrient requirements of the growing bird, in terms of protein, energy, vitamins, and minerals.

If feeds for geese are not available, use a chicken feed formulated for the same age bird. A 15 to 18 percent grower feed is suitable.

Range for Geese

During warm weather, geese can be put on range as early as 1 week of age. A good share of their feed can be forage after 5 to 6 weeks. Geese are better foragers than are ducks. They tend to pick out the young, tender grasses and clover. They reject alfalfa and narrow-leaved tough grasses and select the more succulent clovers and grasses. Geese should not be fed wilted, poor-quality forage.

It takes approximately 1 acre of range to support twenty to forty geese, depending upon their size and the quality of the pasture. Before placing the birds on pasture, make sure the area has not been treated with any chemicals that may be harmful to your birds. A 3-foot (0.9 m) woven-wire fence will confine the geese to the grazing area.

Example Rations for Ducks (in Pounds)

INGREDIENT	STARTER	GROWER	FINISHER	DEVELOPER	LAYER
Yellow corn #2	70.00	73.58	77.25	39.50	59.00
Barley	—	—	—	15.00	15.04
Oats	—	—	—	11.20	—
Soybean meal (48% protein)	18.18	19.70	16.13	12.40	13.95
Alfalfa meal (17% protein)	2.00	—	—	—	—
Fish meal (60% protein)	7.50	—	—	—	—
Meat and bone meal (50% protein)	—	5.00	5.00	—	5.00
Wheat bran	—	—	—	10.00	—
Wheat middlings	—	—	—	8.00	—
D, L- Methionine	0.17	0.22	0.16	0.15	.014
Dicalcium phosphate	0.55	0.28	0.15	1.30	.018
Ground limestone	0.75	0.77	0.86	2.00	6.24
Iodized salt	0.25	0.25	0.25	0.25	0.25
Vitamin-mineral mix	0.20	0.23	0.20	0.20	0.20
Chlortetracycline-50	0.40	—	—	—	—
CALCULATED ANALYSIS					
% protein	20.00	18.30	17.00	15.00	16.00
Metabolizable energy (kcal/lb)	1,400	1,410	1,426	1,200	1,312
% calcium	0.90	0.85	0.80	0.75	2.90
% available phosphorus	0.45	0.40	0.35	0.38	0.35
% lysine	1.12	0.90	0.80	0.70	0.75
% methionine + cystine	0.90	0.80	0.70	0.65	0.65

Note: The ingredients and values used in this chart are representative of those used in commercial feeds formulated to meet the macro- and micronutrient needs of ducks at various ages. If you plan on mixing your own feeds from available ingredients, it is suggested that you consult a poultry nutritionist or your local Extension Poultry Specialist.

Data from North Carolina State University Fact Sheet on Feeding Ducks.

Feed Rations and Feeding Schedule

If good-quality pasture is available, you can restrict the amount of pelleted ration to about 1 to 2 pounds (0.5–1 kg) per goose per week, until the birds are 12 weeks of age. Increase the amount of feed as the supply of good forage decreases, or when the geese reduce their consumption of grass. If you are growing the birds to market as meat, give them pellets on a free-choice basis after they are 12 weeks of age, even though they are on range.

For the first 3 weeks, feed your geese a 20 to 22 percent protein goose starter, preferably in the form of pellets (less waste occurs with pelleted feed). After 3 weeks of age, feed a 15 percent goose grower in the form of pellets. If a specialized grower diet or starter diet for geese is not available, you can use the equivalent chick starter and grower. Again, the diets need not contain coccidiostats or other medications.

Mash or whole grains can be fed alone, or they can be mixed at a 50:50 mash-to-grain ratio. At 3 weeks of age, a mash-to-grain ratio of approximately 60:40 may be used. The proportions should be changed gradually during the growing period until, at market age, the geese are receiving a 40:60 ration of mash to grain. Depending upon the quality and quantity of available range, these ratios may be adjusted up or down slightly. For maximum growth, it is important that mash-and-grain mixtures provide a nutrient intake of approximately 15 percent protein, the same as the all-mash diet.

Wheat, oats, barley, and corn may be used as the whole grains in various mixtures, such as equal parts of wheat and oats. All corn can be substituted when the goslings are 6 weeks old.

Grower-sized insoluble grit should be freely available to geese throughout the growing period.

The Breeding Flock

For the breeding flock, prospective breeders may be selected at about 6 to 7 weeks of age. Choose a few extra at that time to allow for culling in the future. Birds selected for breeders should be placed on a breeder-developer diet, one containing less energy than the starter or grower diets. Often the breeder-developer diet is fed on a restricted basis to prevent the birds from becoming too fat. Birds that are too fat will lay fewer and smaller eggs. When feeding on a restricted basis, have plenty of feeder space or spread the pellets over a wide area to make sure that all of the birds get their share.

To obtain optimum fertility, it is best to feed a special breeder diet. Switch breeder birds to the special diet a month prior to the date of anticipated egg production. If a breeder developer diet is not available, feed a growing ration during this time.

Average Live-Weight, Feed-Consumption, and Feed-Conversion Ratios of White Pekin Ducklings at Different Ages/Mixed Sexes*

AGE IN WEEKS	LIVE WEIGHT	FEED CONSUMPTION		FEED/LB OF WEIGHT GAIN TO DATE
		WEEKLY	CUMULATIVE	
1	0.60	0.50	0.50	0.83
2	1.68	1.64	2.14	1.27
3	2.98	2.55	4.69	1.57
4	4.01	2.55	7.24	1.81
5	5.13	3.27	10.51	2.05
6	6.19	3.57	14.08	2.27
7	6.96	3.87	17.95	2.58
8	7.54	3.39	21.34	2.83

All units given in pounds.
Source: Cornell University.

Selecting and Mating the Breeder Flock

Breeder ducks are usually selected from the spring-hatched birds at 6 to 7 weeks of age. It is possible at that time to differentiate the males from the females by their voices. The females honk and the males belch. (See the section on sex differentiation for further information.) The ratio of drakes to ducks selected is normally one to six. A few extra should be selected to take care of mortality or to permit further selection during the growing period.

Select breeders for vigor, body weight, conformation, and feathering, as well as breed characteristics. Although ducks demonstrate some selectivity in mating, they are essentially polygamous in their mating habits.

Breeding geese prefer to be outdoors and are somewhat different in their mating habits. Some breeds tend to be monogamous, the Canada Goose being an example. Layer breeds mate better in pairs or trios, and the ganders of some of the smaller breeds may accept four or five females. When selecting geese for breeding flocks, select for body size, rate of growth, livability, egg production, fertility, hatchability, and breed characteristics.

Place the geese with their mates at least 1 month prior to the breeding season. The larger breeds of geese mate best in twos or threes or in a ratio of one male to three or four females in large-flock matings. Ganders of some of the lighter breeds will mate satisfactorily with four or five females.

Do not change or mix geese matings from year to year except when the matings are not working or fertile eggs are not being produced. Geese are very slow to mate with new birds, thus making it difficult to make changes in matings or to introduce new males or females into a flock. If matings are changed, keep previously mated birds as far away from each other as possible.

Growth Rate and Feed Consumption of White Chinese, Embden Gosling*

	CONFINEMENT REARED			RANGE REARED		
AGE IN WEEKS	AVG. WT. PER GOSLING	CUMULATIVE FEED CONSUMPTION PER GOSLING	FEED PER LB OF GOSLING TO DATE	AVG. WT. PER GOSLING	CUMULATIVE FEED CONSUMPTION PER GOSLING	FEED PER LB OF GOSLING TO DATE
3	3.3	5.25	1.75	3.3	5.25	1.75
6	8.1	17.32	2.22	7.8	12.60	1.68
9	10.8	34.92	3.32	10.1	19.97	2.03
12	12.3	47.54	3.96	11.6	30.62	2.71
14	12.8	56.09	4.48	11.8	38.10	3.31

*All units given in pounds.
Source: Duck and Goose Raising, Ministry of Agriculture and Food, Ontario.

Housing and Managing the Breeder Flock

Housing for waterfowl does not have to be as tight as that for chickens and turkeys. The pen should be well lighted, well ventilated, and supplied with plenty of dry litter. Waterfowl, and especially geese, prefer to be outside even during severe winter weather. However, in extremely cold weather or in storms, they may prefer some form of shelter. Insulated buildings are not too important. A simple shed, a small house, or an area within an existing barn can be used as a breeder pen. If more than one trio occupies a shelter, divide it to prevent fighting.

Floors and Litter

Dirt floors can be used, but concrete or wood floors are easier to clean and disinfect. Litter materials that are used for brooding are satisfactory for the breeders. Keep the litter dry. Remove wet and caked areas and add dry litter from time to time as required. Floor-space requirements vary depending upon the type of bird and whether or not a yard is available. Ducks in confinement need 5 to 6 square feet (0.5–0.6 sq m) per bird and 3 square feet (0.3 sq m) when a yard is provided. Geese having access to a yard should have 5 square feet (0.5 sq m) per bird in the house. Up to 40 square feet (3.7 sq m) per bird should be provided in the yard.

Ventilation

Leave windows and doors open during the daylight period to allow adequate circulation of air. Good ventilation is also needed at night to prevent overheating during hot periods and to reduce condensation during the winter. Supplementary heat is not necessary.

Nests

The birds will make their own nests in the litter and lay about anywhere in the house. They prefer a somewhat secluded spot. You may provide simple nest boxes in rows along the wall. The nests should be 12 inches (30.5 cm) wide, 18 inches (45.7 cm) deep, and 12 inches (30.5 cm) high. Nests for large geese should be 24 inches (61 cm) square. Some prefer to not partition the nests at all — the tops and fronts of the nests are open. Straw and wood shavings make good nesting materials. Keep the nesting material clean to prevent dirty and soiled eggs. One nest is adequate for every three to five breeders.

Water

Pens can be kept drier if the watering devices are placed outside the house. Ducks and geese can go without water overnight, provided they do not have access to feed. When permitted to eat without available water, they can choke to death on the feed. If the birds are locked in the house overnight with no water, feeders should be empty or closed. If the birds are watered in the house, again, the water supply should be placed above a screened drain.

Yards should slope gently away from the house to provide good drainage. Mud holes and stagnant puddles of water are sources of disease and parasites. Manure will accumulate in the yard, so it should be cleaned occasionally. The frequency of cleaning will depend upon the number of birds and the size of the yard. Large flocks are normally provided 75 square feet (7 sq m) of yard space per bird.

Lighting

Ducks have the tendency to run in circles if startled in the dark. This is a more serious problem with large flocks. Use an all-night light in the house if stampeding is a problem. One 15-watt bulb for each 200 square feet (18.6 sq m) of floor space is adequate to prevent this.

Breeders should not be brought into full production before they reach 7 months of age. If production starts sooner, there will be a problem with small eggs and low hatchability. Fourteen hours of light will stimulate them to produce in the fall and winter months when the days are short.

Females should be lighted 3 weeks before production, and males should be lighted 4 to 5 weeks prior to mating. Prelighting will help fertility. The lamps used for preventing stampeding in duck pens will also stimulate egg production, but larger bulbs give better light stimulation.

Production

Ducks are much better layers than geese: They lay more eggs over a longer period of time. Geese tend to lay every other day but may lay two or more successive days. Daily gathering of eggs will help discourage broodiness (the urge to set) and an accompanying pause in production. Separating broody birds from their mates and confining them with feed and water will sometimes discourage broodiness.

The birds should reach peak production at approximately 5 or 6 weeks after the onset of lay. The rate of production will depend upon the breed, the feeding program, and the management they receive. Older geese make better breeders than young ones. They lay more eggs their second year, and 2-year-old ganders tend to yield better fertility. Geese will lay until they are about 10 years old. Ganders may be kept as breeders for 5 years.

Egg Gathering and Care

Ducks lay most of their eggs during the night and early-morning hours. Collect the eggs early to prevent them from getting soiled and cracked. However, collection of eggs twice daily is recommended, especially in cold weather. Keep the birds in the house each day until the eggs are laid. Eggs laid outside may freeze during severely cold weather, killing the embryos; they can also be badly soiled if laid outside.

Dirty eggs should be washed soon after gathering. The wash water should be warmer than the eggs: The recommended water temperature is 115°F (46°C). If cold water is used, it will cause the egg contents to contract, and bacteria and other organisms may be drawn through the pores of the shell and contaminate the egg contents. This can cause the eggs to explode in the incubator. Wash the eggs in a detergent sanitizer, which may be obtained from an agricultural supply house. Wash for 3 minutes and follow the manufacturer's directions for use.

Badly misshapen, abnormally small or large eggs or eggs that have been cracked should not be saved for incubation. Chances are they will not hatch. Store eggs at a temperature of 55°F (13°C) and a relative humidity of 75 percent. If hatching eggs are to be stored for more than a week prior to incubation, turn them daily. This will prevent the yolks from sticking to the shell membranes, causing a reduction in hatchability. Eggs should be stored in egg filler flats or egg cartons. Prop up one end of the container at an angle of 30 degrees, and each day alternate the propped-up end. Eggs stored for 2 weeks or longer decline in hatchability quite rapidly.

Incubation

The incubation period varies according to the breed. For Mallards, it is about 26½ to 27 days; for Runners, it is 28½ days; and for all other domestic ducks, except Muscovy, it is 28 days. The Muscovy requires

35 days. The Canada and Egyptian geese require 35 days; and all other geese, 30 days.

It is possible to incubate waterfowl eggs by natural means in small flocks. A goose will cover nine or ten eggs and a duck ten to thirteen eggs. While the Muscovy duck is a good setter, most other breeds of ducks are not. If available, use a broody hen, which will cover four to six goose eggs or nine to eleven duck eggs. Turkey hens and Muscovy ducks are larger and better than chicken hens for hatching duck and goose eggs. Ten to twelve goose eggs can be set under a turkey or a duck, depending upon the size of the bird. Hens should be treated for lice before the eggs are set.

Place the nest where the hen won't be disturbed during the incubation period. It may be placed on the ground or inside a building. Provide a source of feed and water nearby for the setting hen. If hens are used to incubate waterfowl eggs, those hatched early should be removed from the nest as soon as they are hatched and placed under a brooder. This will prevent the hen from leaving the nest before completion of the hatch or trampling some of the young if she becomes restless.

Adding Moisture

When chicken or turkey hens are used for setting, it is necessary to add moisture. Sprinkle the eggs with lukewarm water during the incubation period and place the nest and straw on the ground or on a grass-covered turf. This helps increase the moisture. Some growers indicate that if the eggs are lightly sprinkled or dipped in lukewarm water for half a minute daily during the last half of the incubation period, they hatch much better. If ducks or geese have access to water for swimming, no additional moisture is needed. If a setting hen does not turn the eggs, mark them with a crayon or pencil and turn them daily by hand. Usually, the hen will turn them.

Artificial Incubation

Artificial incubation is used extensively for hatching waterfowl eggs. In the large forced-draft machines, a temperature of 99.25 to 99.75°F (37–38°C) is usually used. For the incubation of duck eggs, the wet-bulb thermometer reading should be 86 to 88°F (30–31°C), and during the hatching period, 92 to 94°F (33–34°C). Because the humidity requirements are so much higher for waterfowl eggs than for chicken eggs, the two should not be set in the incubator at the same time.

In the small still-air incubators, the temperature should be 100.5, 101.5, 102.5, and 103°F (38, 38.6, 39.2, 39.4°C), respectively, for the first, second, third, and fourth weeks of incubation. Satisfactory results also may be obtained by running the machine at an even 102°F (39°C) for the entire period. The thermometer reading should be taken at the top of the egg. Eggs in a still-air machine are usually set in a horizontal position and are turned 180 degrees on each turn. Turn the eggs three or four times daily. Moisture requirements are the same for goose and duck eggs.

Candling

The eggs may be candled 4 to 6 days after incubation starts. Candling is done by passing the egg over an electric candling light in a dark room. The living embryo of the egg will appear as a dark spot at the large end of the egg near the air cell. The blood vessels radiate from this dark spot, giving a spiderlike appearance. Embryos that have died will appear as a spot stuck to the shell membrane and no radiating blood vessels will be apparent. Infertile eggs will appear clear. The infertile egg should be removed from the incubator, along with any cracked eggs that are detected.

Eggs are usually candled 3 days prior to hatching. At this time, the fertile eggs will appear dark, except for the large air-cell end. The infertiles will be clear. Discontinue turning the eggs 3 days prior to the end of the incubation period. From that time to hatching, do not disturb the eggs other than to open the incubator to add water.

Hatching

Because the hatching period for waterfowl tends to be spread out over several hours, the large commercial hatcheries frequently take the birds out of the machines as they are hatched and dried. It is necessary to prevent chilling of the newly hatched waterfowl, however, and it may be better for the small producer to keep the incubator closed until the hatch is complete.

Some ducklings appear to need help from the shell. Conditions in small machines are sometimes difficult to control, so this practice may be justified. Birds that are helped from the shell probably should not be used for breeding stock, however, as genetics may be a factor responsible for this condition.

Sex Differentiation of Waterfowl

In general, the determination of sex in ducks is somewhat easier than in geese. Drakes typically have larger bodies and heads, and a soft, throaty quack; and females have a loud, distinct, harsh quack. Also, the main tail feathers are curled forward on the mature drake but not on the duck. With colored breeds of ducks, the males are more brilliantly colored than the females.

Characteristics used to distinguish between the adult male goose and the female goose are the male's longer neck and somewhat coarser and larger head, as well as his larger body. There are differences in voice between sexes, though these are somewhat difficult to determine. The African and Chinese geese have knobs on their heads, and the knobs of the male are larger than those of the female. In immature stock, these differences are hard to distinguish. Even among mature geese, there are cases of oversized females and undersized males. The Pilgrim breed can be sexed from 1 day old to maturity by plumage color: The male is light or white, and the female is a dove gray.

Sexing Goslings and Ducklings

On occasion, it is desirable to be able to sex goslings and ducklings at an early age. Vent sexing of day-old ducklings and goslings is similar. It is done in a warm room using a strong light, so that reproductive organs can be seen easily.

Sexing Adult Geese

To sex an adult goose, hold the bird over a bended knee or on a table — on its back, with the tail pointed away from you. Move the tail end of the bird out over the edge of the knee or the table, so it can be readily bent downward. Insert an index finger into the cloaca about ½ inch and move it in a circular manner several times to enlarge and relax the sphincter muscle, which closes the opening. Then apply pressure directly below and on the sides of the vent to expose the sex organs.

Sexing Day-Old Waterfowl

1. Invert the duckling or gosling, as shown in the illustration, placing the thumb above the vent and the first finger under the back.

2. Place the second and third fingers around the abdomen, holding the legs between them.

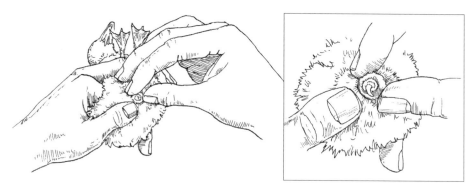

3. Press the thumb and first finger of the other hand together over the vent, parting them slowly so that the vent is extended and the sex organs are exposed.

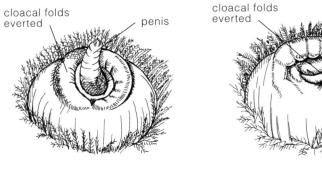

Appearance of the male genital organ

Appearance of the female genital organ

Geese as Weeders

Geese will eat many noxious weeds, and thus have been used to control unwanted vegetation in commercial crops, waterways, and lawns. Geese are good weeders in orchards and for crops such as strawberries, sugar beets, corn, potatoes, mint, cotton, and nursery stock.

There are many benefits in using geese as weeders, including eliminating or reducing the use of herbicides. This is especially important in organic farming, and with the increased interest in organic farming, there is an increasing demand for geese as weeders.

Geese will not compact the soil as heavy machinery and people will, and they work 7 days a week, rain or shine. Their agile necks allow them to pull weeds close to and from within the crop plants, places machines and hoes cannot reach. Also, while weeding your crop, they are adding nitrogen-rich manure all over your field.

You can use geese in a wide variety of crops and situations because their diet includes few broad-leafed plants and favors grasses. Geese will eat young Bermuda grass, Johnson grass, sedge and nut grass, puncture vine, clover, chickweed, horsetail, and many other types of weeds.

For best results, start the goslings as weeders at no more than 6 weeks of age. Provide shade and space waterers throughout the field. Keep the weeder goslings hungry. A light feeding of grain at night is enough and the amount is varied, depending upon the availability of weeds and grass in the crop being weeded.

If placed in a field before the weeds get a hard start, six to eight goslings properly managed can control the weed growth on 1 or possibly 2 acres (0.4 or 0.8 ha) of strawberries. After the fruit begins to ripen, it is best to move the goslings to other crops, or put them on range to fatten for market. A fence 30 to 36 inches (76.2–91.4 cm) high serves to confine the goslings to the area to be weeded. Any pesticide or insecticide used on adjoining tree crops must not contaminate the weeds and grasses to be eaten by the goslings.

Training Weeders

A grower with more than 100 acres (40.5 ha) of herbs has successfully trained geese to eat weeds they normally don't eat. He feeds the young goslings the undesirable weeds, so that they develop a taste for them. He raises a small crop of these weeds in a greenhouse to have them available for the goslings in the spring.

What Breed of Geese to Use as Weeders

Because all geese essentially eat the same vegetation, any breed will work as a weeder goose. White Chinese geese are often used, since they have active foraging habits and their long, agile necks make them effective grazers. Because of their light body weight, they do little damage to the crops they may step on. However, if you want a larger goose for processing at the end of the season, you should probably pick Toulouse or Embden. If you are going for looks, the African geese may be your choice.

Feathers

Duck and goose feathers have value. If properly cared for, they may be a source of extra income, or they may be used at home. The feathers are used mainly by the bedding and clothing industries. It usually requires five ducklings or three goslings to produce 1 pound (0.5 kg) of dry feathers.

Feathers may be sold to specialized feather-processing plants, or a small producer can wash and dry them for home use. Wash by using a soft, lukewarm water with detergent or a little borax and washing soda. After washing, rinse, wring out the moisture, and spread the feathers out to dry.

General Health

Ducks and geese are not normally as prone to disease and parasite problems as some other types of poultry. One reason is that they are hardy birds by nature. Then, too, they are most commonly grown in small flocks and allowed to forage, rather than being kept in close confinement. Where waterfowl are managed in larger numbers and under close confinement conditions, disease and parasites can be of economic consequence.

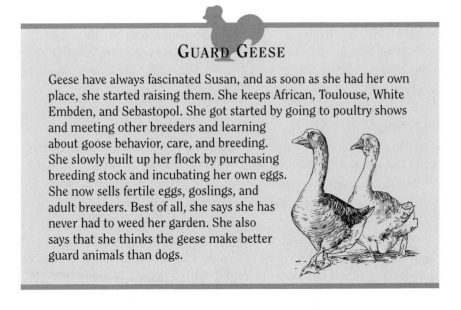

GUARD GEESE

Geese have always fascinated Susan, and as soon as she had her own place, she started raising them. She keeps African, Toulouse, White Embden, and Sebastopol. She got started by going to poultry shows and meeting other breeders and learning about goose behavior, care, and breeding. She slowly built up her flock by purchasing breeding stock and incubating her own eggs. She now sells fertile eggs, goslings, and adult breeders. Best of all, she says she has never had to weed her garden. She also says that she thinks the geese make better guard animals than dogs.

ELEVEN

GAME BIRDS AND OTHER POULTRY

In the past several years, there has been a resurgence of interest in rearing both commercial and ornamental game birds. This is a broad subject and beyond the scope of this book. However, several good references are listed at the end of the chapter to assist you in managing and breeding them.

Many people raise game birds for release in wildlife coverts near their homes, for the gourmet food market, or just for the enjoyment of watching the colorful, unique birds. However, the majority of game birds are reared for restocking private hunting and conservation areas and some public areas. Private hunting preserves are licensed in forty-eight of the United States. The size of any particular game-bird enterprise can range from a small, backyard hobby flock to several thousand birds on a commercial game farm. The common thread is that no matter what size the operation, it takes knowledge and skill to successfully rear the birds.

For many years, the greatest obstacles to rearing ornamental game birds were disease and feeding problems. However, the development of drugs and vaccines has pretty much solved the disease problem. Meeting the individual birds' nutritional requirements is another critical area of management. Although each species of game bird is different, it has been found that, in general, ornamental game birds have similar dietary requirements. There are some exceptions, however, and these will be noted below.

Word of Warning

Before starting a game-bird flock, do more of what you are doing right now — your homework! Get the basic information from the references listed, then contact a successful game-bird breeder in your area and pick his brain. Prepare for your birds in advance. Make sure the habitat you have chosen is the right one for the type of bird you want to rear. Find out about nutrition and possible health issues. Only after you have properly prepared should you place birds in your facility. It is probably best to start with some breeding stock purchased from a reliable, pullorum-free source.

Guinea Fowl

There has been a recent increase in demand for guinea fowl. The meat of a young guinea is tender and of especially fine flavor, resembling that of wild game, and therefore has been substituted for the meat of game birds such as grouse, partridge, quail, and pheasant. Guinea fowl also has many nutritional qualities that make it a worthwhile addition to the diet. It is second only to turkey in low calories, having 134 kilocalories (kcal) per 100 grams (turkey has 109 kcal). The meat is lean and rich in essential fatty acids.

There are other reasons for raising guinea fowl. The bird has been used in protecting the farm flock from intruders because of its loud, harsh cry and its pugnacious disposition. Since one of the main sources of food for wild guineas is insects, they have gained popularity for use in reducing insect populations in gardens and around the home, especially because, unlike chickens, they do not scratch the dirt much and do little damage to the garden. Recently, guineas have been used to reduce the deer-tick population associated with Lyme disease.

Many species of wild guinea fowl are found in Africa and they derive their name from Guinea, on Africa's west coast. There are thirty-eight natural species and subspecies of guinea fowl that are currently recognized and these are divided into four general and seven subspecies. The common domestic guinea fowl are descended from one of these wild species *(Numida meleagris)*. Guineas were domesticated thousands of years ago and were raised for meat by the ancient Greeks and Romans.

Helmeted Guinea

Helmeted
Guinea

There are three principal varieties of helmeted guinea fowl reared in the United States at this time — the Pearl, White, and Lavender. The head and neck are bare, but there may be some wattles. The wattles on the male guinea are much larger than on the female.

Pearl
The Pearl is the most popular variety and the one most people recognize. It has purplish gray plumage regularly dotted, or "pearled," with white spots, and its feathers are often used for ornamental purposes.

White
The next most common variety is the White Guinea, also called African White. The White Guinea has pure-white feathers and its skin is lighter than the two other varieties. These birds are not albino and are the only solid-white bird that hatches solid white and not yellow.

Lavender
Lavender guineas are similar to the Pearl, but with plumage that is light gray or lavender, dotted with white. Crosses of Pearl or Lavender with White produce what is called a "splashed" guinea, the breast and flight feathers being white and the remainder of the plumage pearl or lavender. Selective breeding by fanciers has produced other well-established colors, such as purple, coral blue, buff, and buff dundotte. Newer colors include chocolate brown, sky blue, powder blue, slate, violet, opaline, porcelain, pewter, and bronze. Whatever color pattern your guinea is, the care and management is still the same.

Crested Guinea

The two other varieties of any significance found in the United States are the
Crested and the Vulturine. The Crested
guinea fowl — *Guttera pucherani* — are
chicken-sized birds with small white spots
surrounded by a black spot on a dark chestnut
background. The head is topped with a crown
of curly black feathers. The bare skin of the chin
and throat and down the front of the neck,
around the eyes, and behind the crest is red.
The rest of the bare skin and the neck are
cobalt blue. The sexes look alike. The
flight feathers have white edging and spots
and the outer secondary feathers have broad white edges. Their individual feathers are half black with white spots and half striped.

Crested Guinea

Vulturine

The Vulturine is the only
species in the genus *Acryllium*,
which is to say that there are no
other species that closely resemble
this bird. At 24 inches (60 cm) in
height, the Vulturine is the largest and
most striking member of the family
Numididae. The Vulturine guinea fowl is a
native of sub-Saharan Africa, from
Uganda south into eastern Kenya.
The birds travel in flocks of twenty to
fifty and their voices have been
likened to creaking wagon wheels. Captive-bred Vulturines become tame.
The hens lay a clutch of 8 to 15 eggs. Although they do well in captivity,
they do require protection from colder weather and, where temperatures
reach freezing, they need heated quarters.

Vulturine

Exhibition of Guinea Fowl

Although guineas are not recognized by the American Poultry Association, they are exhibited at many shows and judged for prizes. In judging guineas, the points regarded as most important are good size and uniform color. Presence of white feathers is the most common defect in Pearl and Lavender varieties. At maturity, both male and female guineas range from 3 to 4 pounds (1.4–1.8 kg) in weight.

Basic Management of Guinea Fowl

If you already have other poultry, you will soon discover that guineas are not chickens. They are much more active and not as easily tamed. They seem to retain some of their wild behavior and will remind you of this whenever they get spooked.

Guineas require a dry environment with plenty of room. They are extremely good runners and use this method, rather than flying, to escape predators. Since most people raise guineas with the intention of letting them run loose after reaching adulthood, space is usually not a problem. If you are confining your birds for any length of time, give them as much room as possible outside and a minimum of 2 to 3 square feet (0.2–0.3 sq m) per bird inside. The more room they have, the less likely that they will become overly stressed. Guineas tolerate weather extremes fairly well after they are fully feathered and have reached adult size.

Guineas begin to fly at an early age and can be confined only in covered pens. It is not unusual to find adults roosting 20 to 30 feet (6.1–9.1 m) above the ground, complaining about everything they see. They are strong fliers and will often fly 400 to 500 (122–152.5 m) feet at a time when moving around the farm, especially if startled.

The laying season varies, depending on your latitude and local weather patterns. The Pearl and Purple usually have the longest laying season and the lighter colors have the shortest.

Keet Brooding and Management

Young guineas are called *keets*. They are active birds right from the start and will amaze you with just how lively and fast they are. At hatch and early on, they are smaller than full-size chicks and you should take precautions to contain them, as they will go right through 1-inch (2.5 cm) poultry mesh.

Because guineas are native to dry areas of Africa, the keets do not tolerate being wet. If you are raising keets with a guinea hen or foster hen, keep them in the pen each morning until the dew is off the grass. Guineas make poor mothers and often abandon keets that get separated during daily trips through tall grass. It is not unusual for a hen to lose 75 percent of its brood during the first 2 weeks because of this. It is therefore recommended that you use a chicken as your brood hen for caring for your keets. A large chicken hen can brood as many as twenty-five guinea keets.

Keets can also be brooded on the floor or in cages that are similar to facilities used for chickens or turkeys. Either way works well for guineas. The surface of the floor and feed pans should provide good traction for the newly hatched keets. Set down clean, dry litter prior to the keets' arrival. Guineas have weaker legs than chickens and should never be brooded on newspaper or any slick surface. On smooth surfaces, they have a tendency to go "spraddle-legged" in an extremely short time. Once a leg gets twisted out from under a bird, it is almost impossible to get it to walk normally again. It is recommended placing keets on ¼-inch (0.6 cm) or ⅜-inch (0.9 cm) hardware cloth, because the small holes allow the birds to get traction but do not let them fall through or catch their legs, as might happen with ½-inch (1.3 cm) wire. Start brooding temperatures between 95 and 100°F (35 and 38°C) for the first 2 weeks, and then decrease the temperature by 5 degrees F (3 degrees C) each week. Depending on the temperature in the brooding area and the number of birds you have together, you can usually discontinue the heat after 6 to 8 weeks. As with other brooded birds, watch their behavior to determine if the brooder temperature is too hot or too cold (see page 83).

Adults

Adult guineas require little care and do well on their own. Clean water and a regular laying mash are basically all you need to rear them. They enjoy a little scratch feed mixed in with their feed and scattered on the ground. If your birds are allowed to roam freely, they will eat very little during the summer months. If you have plenty of bugs and seeds, you will start wondering if they are even touching their feed at all.

Feeding Guineas

Keets need a 24 to 26 percent protein ration such as turkey starter or game-bird feed. Use an unmedicated feed to avoid potential problems with keets' being overmedicated. Reduce the protein to about 18 to 20 percent for weeks 5 through 8. After that they will do well on regular laying mash, which is usually 16 percent protein. If you can't find feed with various amounts of protein, mix the higher protein feed with laying mash to get the proper protein mix. The guineas' natural diet consists of a high-protein mix of seeds and insects. If your birds have a large area in which to roam, they will usually get enough to eat on their own; but you can train the birds to stay closer to home by providing supplemental feed in a regular location.

Guineas need a higher protein feed than chickens, but they do quite well on regular poultry mash or crumbles. It is recommended that they be given only mash or crumbles instead of pelleted feed. Adult guineas can usually handle pellets without much difficulty, but due to differences in pellet size, some birds may scratch a lot out of the feeder and waste more than they eat. They will not eat much supplemental feed if they are finding plenty to eat on their own, but it has been found that they favor wheat, milo, and millet and will clean up every kernel. However, give whole or cracked grains, but not too much, only as a treat or a supplement. The protein content is too low and the fat content too high to be of much value. Guineas don't care for the larger grains and will ignore whole corn kernels.

Make sure they have access to clean water. Give keets warm water only. They don't tolerate cold water well.

Sexing Guineas

One of the most-often-asked questions about guineas is how to tell the hens from the cocks. Young guineas cannot be sight-sexed like other poultry or fowl. The hens and cocks look exactly alike, except for the hens of some of the newer colors, which are darker as both keets and adults. The only precise way to tell the sexes apart is to listen for the two-syllable call the hen makes. This has been described as sounding like *buckwheat, buckwheat; put-rock, put-rock;* or *qua-track, qua-track;* and it is the only sound the hen makes that the rooster doesn't. The young birds start making these sounds at 6 to 8 weeks, but some hens do not begin calling until much later.

Incubation

Incubation of guinea eggs takes 26 to 28 days and is similar to incubation of turkey eggs.

Raising Ornamental Quail, Partridge, and Pheasants

Many people raise game birds simply because they like the way they look or act, or just for the simple enjoyment of the company of the birds. Because these are not reared strictly for meat or egg production, they are classified as ornamental birds. All of the species listed are available from aviculturists, with the possible exception of the Barred Quail of Mexico. Rare varieties are hard to find, and that is why they remain rare varieties and command a high price. They are hard to find usually because they are more difficult to rear and breed in captivity. However, if healthy, strong breeding stock is found, they can be successfully reared in captivity.

Quail

In the quail family, there are several inter-esting species: the Barred Quail of Mexico, the Benson Quail (also called the Elegant or Douglas), the Blue Scale (also called the Blue Racer, Blue Quail, Topknot Quail, and Zollin), Bobwhite family of quail, California Valley Quail, the Coturnix Quail family, Gambel's Quail, Mearns Quail, and the Mountain Quail.

Bobwhite Quail

Partridge

In the partridge family, there are forty-seven different species identified. Of the forty-seven, only twelve are reared in captivity, of which about four are not found in the United States. The species available in the United States include the Bamboo, also called the

Mountains Bamboo, the Ferruginous Wood, the Common Hill, the Himalayan Snowcock, the Hungarian, the Madagascar, and the Rock family, which includes the Chukar, Barbary, and Red-Legged partridge. Of these the Chukar are probably the most popular. However, the Hungarian is starting to grow in popularity in game bird-circles.

Chukar Partridge

Pheasants

Pheasants are not native to North America but have become a popular game bird found in many areas of the United States. One of the most familiar breeds is the common or Ring-necked Pheasant. Other popular breeds are the Golden Pheasant, Lady Amherst's Pheasant, Reeve's Pheasant, Silver Pheasant, Jumbo Ring-necked, and Mongolian Pheasant.

Ring-necked Pheasant

Housing and Managing Game Birds

Many successful game-bird breeders have developed their own management methods for breeding and rearing particular game-bird species. However, to go into detail on each species is beyond the scope of this book, and the reader is referred to one of the many books and other resources on game-bird rearing for more detailed descriptions of game-bird management.

Assembly-Line Method

A basic method has been developed by Leland and Melba Hayes, which they call the *assembly-line method*, for the rearing of chicks to breeders. A condensed version is offered here.

As chicks pass through different stages of maturity, their environmental requirements change, and you must meet these requirements. If you are hatching up to twenty chicks at a time, use the box method — a series of boxes, each modified to meet the needs of chicks of different ages. The boxes can be made of regular corrugated cardboard or wood. However, if you have a much larger hatch, you may want to use floor pens and brooders.

Nursery Box

The first box (environment) is the nursery box. It is used to house chicks right from the incubator up to 2 weeks of age. A box of about 15 inches by 20 inches (38.1 x 50.8 cm) and at least 20 inches (50.8 cm) in height works best. It has a cloth floor made of an old towel that can easily be washed, disinfected, and reused. The towel should be changed at least once a day, more if it gets wet or really soiled. A 40- or 60-watt red (or regular bulb painted red) incandescent lamp, depending upon heat requirements and the temperature of the room housing the box, placed about 1 to 2 inches (2.5–5.1 cm) above the chicks' heads near one end of the box, provides the heat. Cover the box with hardware cloth to keep active chicks from jumping out, and cover part of the top with newspaper to control the heat in the box. Watch the chicks carefully. If they start to pant, they are too hot and more heat must be let out. If they are scattered evenly, they have enough heat.

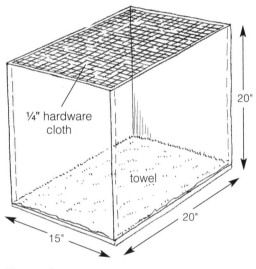

Nursery box

To get the birds to begin feeding, starter feed is scattered on the towel. A 28 to 30 percent protein medicated turkey starter or game-bird starter is preferred. It is all right to let the chicks eat off the floor for the first few days to get them used to eating. After that, feeders should be used. Keep the waterers at the end opposite the heat source. Special molded-plastic water bases that fit standard canning jars (available from game-bird supply companies) are good waterers for quail and other small chicks. If open lids are used, place some rocks or marbles into the lids so the birds will not be able to fall into the water and get wet and chilled. The water should be at room temperature to prevent chilling the birds. Many producers add bacitracin, terramycin, or another water-soluble antibiotic to the water of the baby chicks for the first 2 weeks to help them get a proper start.

Baby Box

After the chicks are 2 weeks old, it is time to move them to the next box, the baby box. It can be about the same size or slightly larger than the nursery box. The difference is a raised floor made out of ¼-inch (0.6 cm) hardware cloth for small breeds and ½-inch (1.3 cm) for the larger partridge and pheasant chicks. This floor should be raised about 2 inches (5.1 cm) from the bottom of the box, allowing the droppings to fall below. You can use litter on the bottom of the box, under the wire, to absorb the manure and moisture. The heat lamp can be moved up as needed and a lower-wattage bulb used to prevent overheating of the chicks. At this age the rocks or marbles can be removed from the waterers.

Chick feeders should be in use by this time. Flat lids with a wire guard to keep the chicks from scratching out the feed and scattering it all over work well. The birds will also be feathered at this time and may fly

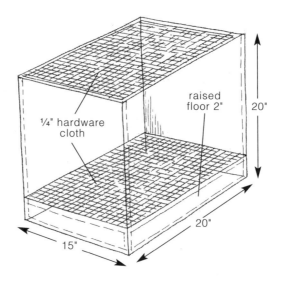

raised floor 2"   20"

¼" hardware cloth

20"

15"

Baby box

out when you remove the wire top to add feed or water; so take care. It is recommended that the wings be clipped each time the birds are moved from one environment to another if they are to be kept in pens higher than three times their height. The primary flight feathers are clipped on one wing.

Juvenile Box

The third box is the juvenile box, which is the same as the baby box, except with ½-inch-mesh hardware wire on the floor. Several of these boxes will be needed to divide the birds into to prevent overcrowding. When the birds' feet are large enough to fit the new wire floor, it is time to move them into the juvenile box. At this time, mixing of different ages of chicks can be done if you follow a few simple rules. It is generally not good to mix species. Never put chicks of other species in with Bobwhites. They will most likely be killed. However, mixing can be done among compatible species if done when you move birds from one box to another. The new environment is new to all the birds and they adapt to one another as well as to the new box. It is for this reason that each box in the "assembly line" be arranged a little differently from the others, so it appears to be a totally new environment to the birds. Another rule is to never add a new batch of chicks to a box already occupied. The current residents of the box will not take kindly to the intruders.

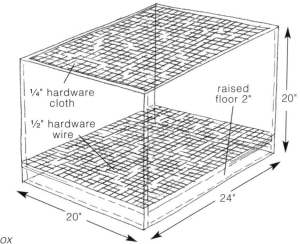

¼" hardware cloth

raised floor 2"

20"

½" hardware wire

24"

20"

Juvenile box

Intermediate Pen

The fourth box, or environment, is the intermediate pen. This is used when the birds have feathered out enough that they do not require supplemental heat at night. It is much larger than the previous boxes, providing the birds with much more room.

Materials

- 2 4' x 8' x ½" (or ¼") plywood
- 80" of 1 x 2 stock
- 4 96" pieces of 2 x 2 stock
- 96" length of 48" wide ½" mesh hardware cloth
- 16" length of 36" wide ¼" mesh hardware cloth
- 1" drywall screws or nails
- ½" staples for attaching hardware cloth

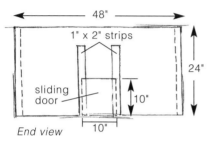

End view

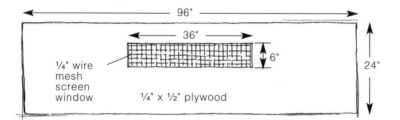

Side view

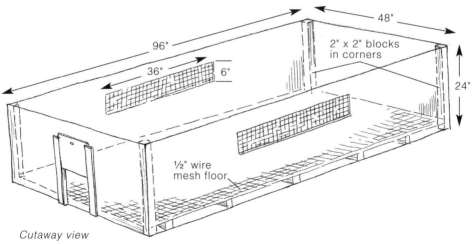

Cutaway view

These pens are made out of ¼- or ½-inch plywood. Construction can be modified, as long as the pen supplies enough room and is usable.

1. Cut two pieces of 4-foot by 8-foot plywood in half (lengthways), making four 2-foot by 8-foot (0.6 x 2.4 m) pieces.

2. Cut one of these in half to make two 2-foot by 4-foot (0.6 x 1.2 m) pieces, which will be the ends of your pen.

3. Cut out 6-inch by 36-inch (15.2 x 91.4 cm) windows in the center of each of the long pieces and cover on the outside with ¼-inch (0.6 cm) wire mesh. This will provide light and air for your birds.

4. Cut a door in one or both ends of the pen. The door can be 8 inches by 8 inches (20.3 x 20.3 cm) or 10 inches by 10 inches (25.4 x 25.4 cm). Cut two strips of 1 x 2 stock twice the height of the door, and then cut a 1-inch-wide (2.5 cm) by ½-inch- thick (or whatever thickness of plywood you are using) strip out of one long edge of each board. These cuts will be the guides for your sliding door.

5. Nail or screw the pen together using 2-inch by 2-inch (5.1 x 5.1 cm) stock in the corners.

6. Place strips of the 2-inch by 2-inch (5.1 x 5.1 cm) stock along the bottom inside edges, so there will be a nailing place for the wire bottom.

7. Cover the bottom with ½-inch (1.3 cm) wire hardware cloth. This is usually sold in 48-inch (121.9 cm) widths, so a piece 8 feet (2.4 m) long should fit the pen area just right. Staple the wire across the bottom on the inside on top of the 2-inch by 2-inch (5.1 x 5.1 cm) strips. This will provide a 2-inch (5.1 cm) space between the bottom of the pen and the ground, or the top of another pen if stacked.

8. If the pen is not placed outside, where it may get wet from rain, then for the top of the pen use heavy-duty cardboard, from discarded shipping cartons used for furniture or appliances. This can be used like a dropping board if these pens are stacked on top of each other, by placing a 2-inch to 3-inch (5.1–7.6 cm) spacer in each corner between stacked pens and slipping newspaper over the cardboard to catch the droppings. The cardboard is also softer than wood and it doesn't hurt the birds' heads as much when they fly up and hit it.

If you plan to use the pen outside, then cover with another sheet of 4-foot by 8-foot plywood.

A light placed in one end over the feeding and drinking area will act as an attraction light at night. It encourages the birds to eat and drink at night and helps keep the birds calmer. This pen can be used until the birds are fully feathered and mature and will hold about forty quail (fewer for partridge and pheasants) as long as care is taken to prevent picking. Some people put fresh alfalfa hay on the bottom to help prevent picking.

When the birds are fully feathered with adult plumage and able to withstand the outside climate without any supplemental heat, they are placed in outside pens. At this time, both wing primaries are clipped to prevent unbalanced flying, which may cause injury to the birds.

Outside Growing Pen

The next step in the assembly-line method is to place the birds in an outside growing pen. This is a large screen pen. If you live in a humid, wet climate, you need to keep the birds on wire off the wet ground. If you live in a dry climate, the birds can be grown on the ground. The dimensions of this outside pen can be almost anything you are comfortable in. Leland Hayes uses one that is 8 feet by 24 feet (2.4 x 7.3 m). The idea is to make the pens longer than wide, so the birds have plenty of room to run up and down and get the exercise they need and to provide a place to retreat if they are threatened. Never put birds into a pen where they can be backed up against a wall. If you do, they will panic and fly up. Provide a shelter in one of the pens for birds to sleep in and where feed and water can be kept out of the weather. Place the feeders and waterers on a platform made of 2 x 4s covered with wire mesh to keep the spilled feed and water off the ground where the birds can get to it.

The birds may not sleep in the shelter and prefer to be in the open. If this is so, put some branches in the open area. It is preferred that the open area have bare ground, since ground wet enough to support vegetation can also support disease agents, earthworms, and other creatures.

Breeders' Pen

The final pen is for breeders only. This is the same pen described above for growing birds, but with different furnishings. Brush, logs, and branches are added to make the environment as natural as possible and to provide plenty of hidden nesting places. If the nesting places can be reached easily from the outside, the eggs can be collected with minimal disturbance. The less the birds are disturbed, the more chance of egg production.

Your own personal experience will guide you in modifying the above assembly-line method to fit your particular species and needs. The size, shape, furnishings, feed, and water are flexible. You just need to meet the needs of the birds you are rearing. The old adage "If it ain't broke, don't fix it" applies best to game-bird management and housing. If you are having success, don't change either unless there is a real need to.

Feeding Game Birds

Game birds, like other poultry, require a balanced diet consisting of all the essential nutrients. This means a balance of protein (amino acids), carbohydrates, fats, minerals, and vitamins, in addition to plenty of fresh water. It is not easy to make a complete balanced diet unless you have access to sources of many types of feedstuffs. Because of this, it is recommended that you purchase feed from a commercial source. There are many good commercial game-bird feeds available today. One main point is to match the protein and energy requirement to the needs of the birds. The weather, size of pen, type and age of bird, and time of year are all factors that influence how much to feed and what protein and energy level should be fed. The following are some general crude protein (CP) and metabolizable energy (ME, kcal/lb) requirements for specific types of birds.

Minimum Protein and Energy Requirements

BIRD/AGE	CP	ME
Bobwhite		
0–4 wk	28%	1,350
4–9 wk	24%	1,375
9–18 wk	18%	1,375
Breeders	19%	1,250
Coturnix		
0–3 wk	25%	1,350
3–6 wk	23%	1,350
6+ wk	18%	1,350
Breeders	18%	1,250
Pheasant		
0–4 wk	30%	1,150
4–9 wk	24%	1,150
9–18 wk	18%	1,240
Breeders	19%	1,150

CP = crude protein; ME = metabolizable energy requirements in kcal/lb.

It is recommended that commercially formulated and mixed feeds be used, if possible, rather than grains fed alone, which may not supply all the required nutrients. A good turkey starter followed by turkey grower rations usually will suffice for starting your birds and growing them to maturity if other feeds are not available.

Starter Feed

During the first 5 to 6 weeks of life, quail, partridges, and pheasants should receive only high-protein (28–30%) starter feed, with no additional grain. For best results and the least feed wastage, the starter feed should be in mash form the first 2 weeks, small crumble form weeks 3 and 4, and small pellets thereafter.

Grain Feeding

Feeding your birds just a little grain by hand every day gives you a good chance to look the birds over and provides some personal contact. A little extra corn during periods of stress, such as during cold weather or when the birds are moved, will help provide the energy needed for extra heat or to get them through the stressful period.

When feeding grain by hand, you can also check your birds for signs of disease and study their habits. It will also tame your birds somewhat and make them easier to catch and handle. When whole-grain feeding is begun and grain is fed in combination with pellets, be sure to provide grit at all times in separate hoppers. If cracked, crushed, or milled grains are utilized, grit is not necessary.

Don't overfeed grain or your game birds may not get a proper balance of the essential nutrients contained in the complete feed. After week 10, grain may be fed in the morning and the evening.

Some breeders will supplement the feed with fresh fruits and vegetables. Quail and partridge like apples. Chopped carrots are also readily eaten by some species. Fresh greens are good for the birds. These foods provide extra nourishment and keep them occupied for a while.

Many breeders also like to feed mealworms to their birds. These can be grown by you or purchased from pet stores or from sources listed in the *Game Bird and Conservationist's Gazette* or *Wildlife Harvest*.

Quail-Feed Consumption

Quail nutritionists say that quail will consume an average of 0.8 pounds (0.4 kg) of starter feed from 0 to 6 weeks of age and 0.25 pound (0.11 kg) per bird per week of flight conditioner from 6 to 16 weeks. Total feed consumption from 1 day through 8 weeks would be 1.3 to 1.5 pounds (0.6–0.7 kg) per bird. From 8 to 16 weeks, it is estimated that quail may consume 2.5 to 3.0 pounds (1.1–1.4 kg) per bird.

Quail breeder-layers may consume 0.3 pound (0.14 kg) of ration per bird per week (1 month before start of egg production throughout the laying season). Quail typically consume 0.25 pound (0.11 kg) of maintenance ration per bird per week after 16 weeks.

Game-Bird Vices

Almost every species of animal kept in captivity, including game birds, develops some bad trait or vice that must be dealt with. The best way to deal with vices is to prevent them before they begin. An experienced breeder will try to adjust the conditions to suit the birds rather than forcing the birds to adjust to the conditions the breeder dictates.

Still, a number of vices can develop in game-bird flocks. Although it is not always easy to determine the cause, several factors are believed to be involved, including stresses brought on by lack of adequate floor, feeder, or water space; overheating; high light intensity; and poor nutrition.

Cannibalism

One thing that can easily creep up on the bird raiser is cannibalism. Game birds are naturally aggressive, to help them survive in the wild. Cannibalism usually begins as feather picking when the birds are only a few weeks old. Poultry scientists have found that any number of stress factors can trigger picking and lead to more serious cannibalistic outbreaks and behavior. Some of these stress factors include overcrowding, high light intensity, inadequate ventilation, too-high temperature, too-high humidity, insufficient feeding and watering space per bird, external parasites, trace-nutrient and/or salt deficiencies, nervous and excitable birds (normally inherited traits characteristic of the flight strains of ringnecks), and boredom.

Early Signs

Watch for the early signs of cannibalism:

- Vent pecking of birds at their own vent or others'. Vent pecking can also be a sign of enteric infections.
- Feather pulling is frequently seen in birds too closely confined, especially young flocks crowded during brooding periods.
- Toe pecking is common among juvenile birds and is reported to begin due to hunger.
- Head pecking usually follows an injury.
- Nose pecking is seen often in 2- to 7-week-old quail.

Prevention and Cure

Preventing cannibalism is far easier and cheaper than curing it. Provide ample space for birds at feeders and waterers and in pens to help alleviate cannibalism. Provide good, nutritious feed, lots of leafy green alfalfa, and fresh, cool water at all times. Keep the brooder area well ventilated and under subdued light. Red lights and a dark box also seem to help. Providing cover in the outside growing pens will give birds a place to escape, if necessary. Beak trimming helps reduce the injuries from pecking.

Stemmy alfalfa hay on the floor of the pens seems to discourage toe picking. The hay stems break up the outline of the toes and make it hard for the birds to distinguish toes from alfalfa.

A head of lettuce or cabbage hung on a string gives the birds something else to pick at and seems to help reduce cannibalism. If all else fails, specs or peepers can be fitted to the birds.

Egg Eating

Some game birds are egg eaters. This is a difficult habit to break. Some breeders have tried using stones shaped and colored like eggs. The birds peck at them, can't break them, then ignore real eggs. Another remedy that may work is to coat a few eggs with oil of ipecac, obtained from your pharmacist. After the birds eat the eggs, they will regurgitate, something not normal for them. They remember being sick after eating eggs and never break and eat them again. The same technique may even work for feather picking by putting some on the tail and other feathers of birds being pecked.   1 tyvalosin

Game-Bird Health 3. Thiabendazole

The best way to maintain the health of your game birds is with prevention. As stated earlier, biosecurity is your best weapon, and the two major aspects of this are sanitation and isolation. The same elements of disease that cause problems in other birds apply to game birds. (See chapter 13 on flock health for further information.)

Many of the diseases of poultry, such as pullorum, paratyphoid, streptococcus, aspergillosis, coccidiosis, and fowl pox, are common to all game birds.

Some diseases commonly diagnosed in pheasants are mycoplasmosis, paratyphoid, botulism, marble spleen, internal parasites, coccidiosis, Newcastle disease, and gapeworm.

Quail are commonly diagnosed with ulcerative enteritis, enteritis, blackhead, bronchitis, colibacillosis, and internal parasites.

Chukars are diagnosed with histomoniasis (blackhead), mycoplasmosis, fowl pox, paratyphoid, amebic dysentery, mycosis, and internal parasites.

A few diseases that present particular problems in game birds are discussed below.

Ulcerative Enteritis

Ulcerative enteritis, also called quail disease, is the most common disease of quail. It is a bacterial infection of the intestines, and mortality of young birds may be as high as 100 percent if the disease is not controlled. The addition of zinc bacitracin or bacitracin methylene disalicylate at 20 grams per ton (907.2 kg) of complete feed helps control this disease. Streptomycin can also be used. Prevention is best accomplished by good sanitary conditions and raising young birds on wire.

Quail Bronchitis

Quail bronchitis can quickly destroy a commercial quail business. It is caused by an avian adenovirus and appears to affect mainly Bobwhite Quail. It is highly contagious and spreads quickly through the flock. It affects the respiratory and intestinal tracts. It usually starts in birds 2 to 3 weeks old and works back with subsequent hatches to where signs start at 7 to 10 days. Mortality can run from 40 to 100 percent if the disease is not treated quickly. Tylosin helps in most cases.

Gapeworm

Gapeworm primarily affects young pheasants and turkeys. Gapeworm larvae can be picked up from contaminated soil directly or through earthworms infected with the larvae eaten by the bird. They live in the trachea, bronchi, and bronchioles (lungs) of birds, causing them to gasp for air, "gapes." Many infected birds die from a lack of oxygen. Gapeworm is primarily a problem in young birds reared on the ground. Keeping the ground dry and tilling the soil help control the earthworms and the gapeworm larvae. Thiabendazole has been the drug of choice, but new medications may soon be approved for use in gape birds. Extra vitamin A also helps.

Coccidiosis

Coccidiosis affects all avian species. It usually occurs in birds reared on litter or dirt. Wet, humid conditions are ideal for the propagation of this protozoan parasite. All chicks should be started on medicated feeds to help build immunity. If an outbreak occurs, amprolium in the water is the recommended treatment.

Pullorum

Pullorum is found around the world in just about all poultry-producing areas. The disease is caused by the bacterium *Salmonella pullorum*, and is commonly spread by egg transmission from one generation to another, and then from infected chicks to noninfected chicks. There is no cure for pullorum and it is recommend that birds only from certified pullorum-free flocks be purchased. Your local Extension specialist has the latest National Poultry Improvement Plan (NPIP) listing of pullorum-free hatcheries and breeders.

If you have sick birds or birds showing signs of disease, it is best to take a few with the symptoms to the nearest diagnostic laboratory and have them properly diagnosed. If birds die, put them in plastic bags, keep them cool but not frozen, and take them to the lab for necropsy. A few dollars spent on diagnosis will save you a lot more later. Always seek a proper diagnosis by a trained professional.

Controlling Ticks and Bugs

Jill and her husband, Ray, who live in southern New England, were worried about ticks and Lyme disease. They had heard that guinea fowl eat ticks and called the Extension Poultry Specialist to find out more about it. They were told that guinea fowl eat all kinds of bugs and vegetation and, if properly managed, would reduce the tick population on their property; but they were cautioned that even guinea fowl might not get them all and that they must still check themselves for ticks.

Jill and Ray purchased four guinea hens as chicks and hand-raised them so they would be a bit tamer and would stick around the property. During the summer, they restrict the birds' feed to keep them hungry and allow them to roam the property, eating all the bugs they want. They have been very happy with the results and enjoy watching these interesting but strange birds run around and fly to tree branches. At night, Jill and Ray blow a whistle. The birds come running to be fed and are then locked in their coop, protected from predators.

Jill and Ray have spread the word about the guineas, and they are becoming popular birds in many parts of the eastern United States.

Additional Resources

For more information on game birds or poultry in general, check out some of the material listed below.

Web Sites

www.poultryconnection.com
   This is a site filled with poultry information. It is a large site with many pages and is well organized.

home.att.net/~DanCowell/personal.html
   A site loader with information and great photos of game birds, waterfowl, peafowl, and guinea fowl.

www.gfba.org
   A couple of sites for the guinea-fowl enthusiast.

www.manzanodragon.com/manzanovalley/cyberindex.html
   This site contains numerous articles on health and management of poultry and game birds.

Books and Magazines

Ferguson, Jeannette S. *Gardening with Guineas: A Step-by-Step Guide to Raising Guinea Fowl on a Small Scale.* Waynesville, OH: FFE Media, 1999.
   Contains information on raising guineas on a small scale. Much of the information is from the author's firsthand experience with raising guineas.

Game Bird and Conservationist's Gazette. Salt Lake City, UT: Allen Publishing Co.
   The oldest game-bird magazine in the world. Contains articles on breeding and rearing all types of game birds. A must for the true enthusiast.

Hayes, Leland B. *Upland Game Birds, Their Breeding and Care.* Valley Center, CA: Leland Hayes' Gamebird Publications, 1997.
   Written by one of the recognized experts in the field of game birds, this book is one of the best. It will soon be available on CD-ROM.

Hayes, Leland, and Melba L. Hayes. *Raising Game Birds*. Valley Center, CA: Leland B. Hayes, 1987.
  Practical information for the beginning and experienced game-bird breeder on how to raise ornamental quail, partridges, and pheasants. Leland Hayes is a noted expert in the field.

Johnsgard, Paul A. *The Pheasants of the World: Biology and Natural History*, 2nd ed. Washington, DC: Smithsonian Institution Press, 1999.
  Paul Johnsgard is one of the most prolific authors on the subject of game birds and waterfowl. This book is 398 pages of power-packed information about pheasants. The first part of the book covers the comparative biology of the pheasants, including growth, social behavior, ecology, mating, and reproductive biology. There's also a chapter on aviculture and conservation. Following this are taxonomic keys and species accounts.

Larson, Jean A. *Raising Quail, Partridge, Pheasant, Bobwhite and Ostriches: January 1987–1992*. Beltsville, MD: National Agricultural Library, 1992.
  You can access the National Agricultural Library on the Internet at www.nalusda.gov and request this bibliography of articles written about game birds. Articles on all aspects of managing and rearing game birds are listed.

Mullin, John, and Peggy Mullin Boehmer. *Game Bird Propagation: The Wildlife Harvest System*. Goose Lake, IA: John M. Mullin, 1978.
  Available through Wildlife Harvest Magazine, Goose Lake, Iowa, this 320-page book is a practical guide to raising game birds for both the novice and the professional. Species covered are Bobwhite Quail, Coturnix Quail, Chuckar, Red-legged Partridge, Hungarian Partridge, Ring-necked Pheasants, wild turkeys, geese, and Mallard ducks. This book is a must for commercial raisers; the hobbyist will find it equally helpful.

Sainsbury, David. *Poultry Health and Management: Chickens, Turkeys, Ducks, Geese and Quail*, 4th ed. Malden, MA: Blackwell Science, 1999.
  A good book on health and diseases of various types of poultry; recommended by many game-bird breeders.

Scott, Thomas G. *Bobwhite Thesaurus*. Edgefield, SC: International Quail Foundation, 1985.
  This book contains information both the Bobwhite enthusiast and casual breeder will find fascinating.

United States Department of Agriculture. *Raising Guinea Fowl*. Leaflet No. 519. Washington, DC: Government Printing Office, 1976.
  Raising Guinea Fowl can be obtained from your local Extension Poultry Specialist or from the Government Printing Office. This is a revision of the classic Farmers Bulletin No. 1391, first published in 1924, which was a revision of the original Farmers Bulletin No. 858 published in 1917 and authored by Andrew S. Weiant.

Van Hoesen, Roy W., and Loyl Stromberg. *Guinea Fowl*. Pine River, MN: Strombergs, 1998.
  This is a republication of an old book by the authors and is available from Stromberg Chicks and Game Birds Unlimited. It is a very basic guide to some of the breeders of guinea fowl and some basic guinea fowl management.

Wildlife Harvest Magazine. John M. Mullin, Editor and Publisher. Goose Lake, IA 52750.
  A good magazine with articles about game-bird management, health and nutrition, and wildlife conservation.

TWELVE

HOME PROCESSING OF EGGS AND POULTRY

Eggs, as they come from the nest, vary in several respects. They vary in size, shape, cleanliness, shell texture, and interior quality, and some will be cracked and checked. If the right management, the right environmental conditions, and the proper feed have been provided to a healthy flock, the majority of the eggs will be of good quality. And if a family's consumption of eggs keeps up with the birds' production, there won't be too many problems with quality differentiation or quality preservation. On occasion, however, some flocks produce more eggs than can be consumed by a family, and some of them have to be sold. Eggs are sold based on size and quality; thus, it is necessary to understand a little bit about egg-quality determination and egg-size requirements.

Egg Quality

The quality of an egg never improves after it is laid. Quality diminishes with time and the rate depends upon how the eggs are treated. To best understand what happens to egg quality, a knowledge of the structure and the various parts of an egg is essential.

Parts of an Egg

There are four basic parts of an egg — the *shell*, the *shell membranes*, the *albumen*, and the *yolk*. Fastened to the shell are two membranes. It is between these two membranes that the air cell is formed, usually at the large end of the egg. When the egg is laid, this air cell is nonexistent; but with the cooling and contraction of the egg contents, the air cell begins to appear a short time after the egg is laid. Since the shell is porous, moisture can escape from the egg, so the air cell enlarges as the egg is held for a period of time. The rate at which the air cell increases in size depends upon age, shell texture, and the holding conditions — primarily temperature and humidity. The size of the air cell is not a factor when determining the broken-out quality of the egg, but it is a factor when candling for quality. (See illustration on page 128.)

A relative humidity of 75 percent in the cooler will keep evaporation to a minimum. The yolk of a high-quality egg is near the center of the egg. On the surface of the yolk is a small white spot known as the *germinal disc*, *germ cell*, or *blastodisc* (true egg cell or ovum). It is at this part of the egg that fertilization takes place. After fertilization, the blastodisc becomes the *blastoderm* and starts to develop as the egg travels down the hen's reproductive tract. A trained eye can look at freshly laid eggs and determine whether there is a blastodisc or a blastoderm. This is because the blastoderm has an outer ring of opaque cells called the *area opaca* that is attached to the yolk and an inner portion of semitransparent cells called the *area pellucida* that is detached from the yolk. So the blastoderm, or fertilized egg, has a doughnutlike appearance, while the unfertilized egg has a disorganized, bubbly appearance. The yolk and the germ cell are enclosed in a thin membrane known as the *vitelline membrane*.

The *albumen*, or white of the egg, is made up of alternate layers of thick and thin albumen. Surrounding the yolk is a thin layer of dense albumen called the *chalaziferous layer*. From this layer of dense albumen extend the fibrous or cordlike structures called *chalazae*. These extend toward the ends of the egg and are anchored in another layer of thick albumen. The chalazae tend to keep the yolk near the center of the egg. They are the darker-looking little squiggles you see on either side of the yolk of a broken-out egg.

Interior Quality

Between the chalaziferous and outer layer of thick albumen, there is an inner layer of thin albumen. If an egg is broken out and the thick albumen surface is cut, some of the thin albumen can be seen merging with the thick.

After the egg is laid, a number of changes take place as far as the interior quality is concerned. The rate at which this occurs depends upon how the egg is handled. As mentioned earlier, when an egg is laid, there is no air cell; but as the contents cool and contract, the air cell develops. A further increase in the size of the air cell is dependent upon the rate of evaporation of moisture from the egg. The rate depends upon the thickness and texture of the shell, as well as the temperature, relative humidity of the storage area, and length of the storage period. Air-cell size is a factor in grading eggs but actually has no significance as far as eating quality is concerned.

During the holding period, a change takes place in the thick albumen. It gradually breaks down into thin or watery albumen. If the egg is stored long enough, eventually there will be no evidence of thick albumen. As a result of these changes, when older eggs are broken out into a pan, the contents seem to spread over a large area and the yolk has a flattened and enlarged appearance. Most individuals frown on this type of egg because it is a sign of poor quality. An egg of high quality, when broken out, will have a large amount of thick albumen adhering to the yolk. The yolk will be upstanding and practically spherical in shape and the egg will occupy a much smaller area in the pan. If a high-quality egg that is hard-cooked is cut in two, the yolk will be well centered and the air cell small. With a low-quality egg, the yolk may be close to or touching the shell membrane, and the air cell quite large.

Egg quality can be determined relatively easily with the use of a candling light, as discussed earlier (see page 129).

The standards of quality for individual shell eggs are set up by the U.S. Department of Agriculture.

Blood or Meat Spots

It should be noted that blood spots or meat spots are permissible in B-grade eggs, provided the single defect, or the total of several defects, is not more than ⅛ inch (0.3 cm). Eggs containing blood or meat spots not within these tolerances are classified as inedible eggs.

Blood spots in the egg are usually found on the surface of the yolk. They may vary in size from a small speck to a large clot, with some of the blood diffused throughout the albumen. Blood spots are caused by the rupture of one or more small blood vessels in the yolk follicle at the time of ovulation or in the oviduct during egg formation.

Meat spots are either blood spots that have changed in color due to chemical action or tissue that has sloughed off the oviduct of the hen during egg formation. Possibly 2 percent of eggs produced contain blood or meat spots.

Summary of U.S. Standards for Quality of Individual Shell Eggs

QUALITY FACTOR	SPECIFICATIONS FOR EACH QUALITY FACTOR		
	AA QUALITY	A QUALITY	B QUALITY
Shell	Clean; unbroken; practically normal	Clean; unbroken; practically normal	Clean to slightly stained; unbroken; may be slightly abnormal
Air cell	⅛ inch or less in depth; may show unlimited movement and be free or bubbly	3⁄16 inch or less in depth; may show unlimited movement and be free or bubbly	More than 3⁄16 inch in depth; may show unlimited movement and be free or bubbly
White	Clear; firm	Clear; may be reasonably firm	Weak and watery; small blood or meat spots present*
Yolk	Outline slightly defined; practically free from defects	Outline may be fairly well defined; practically free from defects	Outline may be plainly visible; may be enlarged and flattened; may show clearly visible germ development but no blood; may show other serious defects

*If they are small (aggregating not more than ⅛ inch in diameter).

Note: *Eggs that fail to meet the requirements of the above consumer grade are classified as* restricted eggs, *and except for "checks" or "cracks" can be sold to consumers only within the specified tolerances stated in the above grades. Dirties, leakers, inedibles, losses, checks, and incubator rejects are classified under the restricted categories.*

Exterior Quality

In addition to the interior-quality factors we have discussed, there are exterior-quality factors that must be considered. The more important exterior egg-quality factors include condition or soundness of the shell and cleanliness. Leakers, dented cracks, and eggs with rough or thin shells can be detected during processing without candling. Blind checks or hairline cracks can be seen during candling. Dirty and stained eggs should not be marketed.

Weight classes of shell eggs (see table) are set up by the U.S. Department of Agriculture. The system uses the weight in ounces per dozen. For example, a single egg weighing 2 ounces (56.7 g) is called a 24-ounce (680.4 g) egg, because a dozen of this size egg weighs 24 ounces (680.4 g). Uniformity of size in a market pack of eggs is important, and there should be no more than a 3-ounce (85 g) variation from one dozen to another. Small individual egg scales or larger automated scales are available for sizing eggs.

U.S. Weight Classes for Consumer Grades or Shell Eggs

SIZE OR WEIGHT CLASS	MINIMUM NET WEIGHT PER DOZEN (OZ)	MINIMUM NET WEIGHT PER 30-DOZEN CASE (LB)	MINIMUM FOR INDIVIDUAL EGGS AT RATE PER DOZEN (OZ)
Jumbo	30	56	29
Extra large	27	50½	26
Large	24	45	23
Medium	21	39½	20
Small	18	34	17
Peewee	15	28	—

Source: Egg Grading Manual Agricultural Handbook No. 75.

Care of Eggs on the Farm

As stated earlier, immediately after the egg is laid, the quality begins to deteriorate. The sooner the egg is removed from the nest, cleaned, cooled, and packed, the better it is. Some management recommendations that will lead to higher-quality eggs are the following:

1. Keep the birds confined.
2. Gather eggs frequently, at least three times a day.
3. If eggs must be cleaned, clean them immediately after gathering.
4. Dry-clean slightly dirty eggs.
5. Cool eggs as quickly as possible.
6. Maintain an egg-storage-room temperature of 45°F (7°C) and a relative humidity of 75 percent.

Processing Eggs

If eggs must be washed, wash them at a temperature of 110 to 115°F (43–46°C) in clean water with an approved detergent sanitizer.

Washing time should be no more than 3 minutes. If the eggs are rinsed following the washing, use water containing a detergent sanitizer, then dry and cool. Slightly soiled eggs may be dry-cleaned with an abrasive material such as emery cloth, fine sandpaper on a hand buffer, or steel wool.

On commercial egg farms, the washing and sizing of eggs is done with large automated equipment. Smaller operations may use immersion washers; however, USDA-inspected eggs are not allowed to be cleaned in immersion washers, but some states still allow this under inspection. The eggs are immersed in warm detergent sanitizer, which is agitated to clean the eggs.

The main thing to remember is that eggs are perishable: They don't improve with age. It's important to cool them quickly and keep them cool. It's also important to realize that eggs will absorb odors and pick up off flavors. Musty odors, onions, and other vegetables can spoil the taste. Eggs should be packed in clean egg cartons, large end up, to prevent off flavors and maintain fresh quality.

Shelf Life

A question frequently asked is, "How long can you keep eggs and still have them usable?" For years, when egg production was seasonal, eggs were stored for several months in cold storage. They weren't all AA quality when they came out of storage, but most were edible.

The shelf life of eggs can be enhanced when the shells are oil treated. A special colorless and odorless mineral oil is available in aerosol spray cans. Usually, unwashed eggs keep better than washed eggs. It is not unreasonable to expect eggs to keep in your refrigerator for 4 or 5 weeks.

Tips for Safe Egg Handling

With the increased interest in food safety and the concern about salmonella and eggs, some good tips to remember for safe handling of eggs are the following:

◆ Refrigeration, the first step in proper egg handling, retards bacterial growth and maintains the quality of eggs. At retail, buy eggs only from refrigerated cases and refrigerate them in their cartons on an inside shelf as soon as possible after purchase. Today's home refrigerators are designed to maintain a temperature of 40°F (4°C) or below, a satisfactory temperature for eggs and other perishable foods.

◆ Keep shell eggs, broken-out eggs, or egg mixtures refrigerated before and after cooking.

◆ Do not leave eggs in any form at room temperature for more than 2 hours, including preparation and serving. Promptly after serving, refrigerate leftovers in shallow containers so they will cool quickly.

◆ For picnics or outdoor parties, pack cold egg dishes with ice or commercial coolant in an insulated cooler or bag.

◆ Cleanliness of hands, utensils, and work surfaces is essential in preventing cross-contamination.

◆ Use only clean, unbroken eggs. Discard dirty or broken ones.

◆ Avoid mixing the shell with the egg's contents.

For more information, visit the American Egg Board's Web site at www.aeb.org/

Poultry Meat Processing

At best, the job of processing poultry meat is a messy one. Ideally, there should be two rooms available for the procedure: one for killing and picking the birds and the other for finishing, eviscerating, and packaging. If this is not possible, do the killing and plucking in one operation. Clean the room, then draw and package the birds as a second step. This procedure will make the whole operation far more sanitary.

Care before Killing

Remove feed from the birds 6 to 12 hours before they are to be killed. This will give ample time for the crop and intestines to empty. Removing feed from the birds makes the job of eviscerating much cleaner and easier. Remove them from the pen and put them into coopers containing wire or slat bottoms, so that they do not gain access to feed, litter, feathers, or manure.

Use care in catching and handling the birds to prevent bruising. They should be caught by the shanks and not be permitted to flap their wings against equipment or other hard surfaces. This will help prevent bruising and poor dressed appearance. After the birds are caught, keep them in a comfortable, well-ventilated place prior to killing. Overheating or lack of oxygen can cause poor bleeding, resulting in bluish, discolored carcasses.

Equipment Required

The job of processing the birds is made easier with the proper equipment, much of which you might already have on hand or can make yourself. The basic equipment you need are shackles or killing cones, knives, a scalding tank, a thermometer, and a weight or weighted blood cup.

Shackles or Killing Cones

When only a few birds are to be dressed, a shackle can be made from a cord with a block of wood 2" x 2" (5 x 5 cm) square, attached to the lower end. Make a half hitch around both legs and suspend the bird upside down. The block of wood prevents the cord from pulling through. Commercial dressing plants use wire shackles that hold the legs apart and

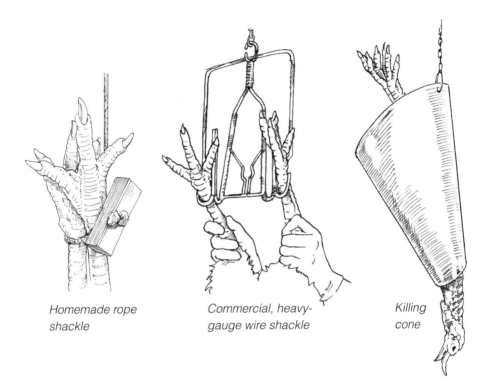

Homemade rope shackle

Commercial, heavy-gauge wire shackle

Killing cone

make for easier plucking. Some producers make their own shackles out of heavy-gauge wire. Others prefer to use killing cones, which are similar to funnels. The bird is put down into the funnel with its head protruding through the lower end. This restrains the bird and prevents some of the struggling that often leads to bruising or broken bones.

Knives

About any type of knife is satisfactory for dressing poultry. There are specialized knives for killing, for boning, and for pinning. Six-inch (15 cm) boning knives work well. For braining birds, use a thin sticking knife.

Scalding Tank

If only a few birds are to be dressed, a 10- to 20-gallon (37.8–75.7 L) garbage can — or any other clean container of suitable size — is satisfactory. If considerable dressing is done, a thermostatically controlled scalding vat is preferred.

Thermometer

Accurate temperatures are important for certain types of scalding. You should have a good, rugged dairy thermometer or some other type of floating thermometer that accurately registers temperatures of 120 to 212°F (49–100°C).

Weight or Weighted Blood Cup

A weight or weighted blood cup attached to the bird's lower beak will prevent it from struggling and splashing blood around. The weight may be made from a window weight and attached to the lower beak by means of a sharp hook. The blood cup is not used when killing funnels are available.

Weighted blood cup

Killing Methods

Suspend the bird by its feet with a shackle or place it in a killing cone. Hold the head with one hand and pull down for slight tension to steady the bird. With a sharp knife, sever the jugular vein by cutting into the neck just in back of the mandibles. Do this by inserting the knife into the neck close to the neck bone, turning the knife outward, and severing the jugular (illustration A). You may also do it by cutting from the outside. Another method is to cut the jugular vein from inside the mouth. With this method, hang the bird with the breast toward you. Hold the head firmly with your thumb and first finger at the earlobes. A slight pull with pressure will cause the beak to open. Insert the knife into the mouth, so that the point can be felt just behind the left earlobe as you face it. (It will be opposite for left-handers.) With a slight pressure

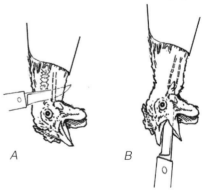

A   *B*

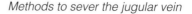

Methods to sever the jugular vein

and drawing outward toward the opposite corner of the mouth, cut the jugular vein at the junction of the connecting vein running across the back of the throat (illustration B, page 248). *Warning:* Hold the bird's head in such a way that your fingers are safe.

During the bleeding process, restrain the bird by holding the head until the bleeding and flopping stop, or attach a weight to the lower beak. Do not grasp the wings or legs tightly, to avoid restricting the flow of blood from these parts. A poor-appearing dressed carcass will result if the bleeding is incomplete.

Debraining

Debraining loosens the feathers, making it easier to pluck the birds. Do this after the jugular vein is cut. Debraining is done when the birds are to be dry-picked, but may also be done when they are to be semiscalded to make the removal of feathers easier.

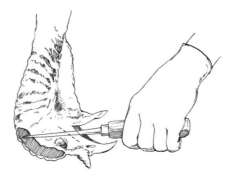

Proper knife insertion for debraining

Insert the knife through the groove or cleft in the roof of the mouth and push through to the rear of the skull to pierce the rear lobe of the brain. Then give the knife a quarter turn. This kills the bird and loosens the feathers. A characteristic squawk and shudder indicates a good stick. If the front portions of the brain are pierced, it may cause the feathers to tighten. This procedure requires considerable practice before you become proficient.

Picking

In general, there are four methods of removing feathers from birds. They are the hard scald, the subscald, the semiscald, and dry picking. The principle behind scalding is to relax the muscles that hold the feathers.

Hard Scald or Full Scald

This method is probably used more commonly on the farm than any other. The hard scald uses 160 to 180°F (71–82°C) temperatures for 30 to 60 seconds. After the bird is sloshed up and down in water at this temperature, the feathers are easily removed. With this method, the scalding time depends upon the temperature of the water and the age of the birds. Scald only long enough to allow the feathers to be pulled easily. This method would conceivably be used only for older birds and possibly waterfowl. Hard scalding makes for fast, easy picking but destroys the protective covering of the skin. The hard scald causes a dark, crusty, blotchy appearance and results in poorer keeping quality.

The Subscald

The subscald uses a temperature of 138 to 140°F (59–60°C) for 30 to 75 seconds. This method causes a breakdown of the outer layer of skin, but the flesh is not affected as in hard scalding. The main advantage of the subscald is the easy removal of feathers and a uniform skin color. The skin surface tends to be moist and sticky, however, and will discolor if not kept wet and covered. This method is frequently used for turkeys and waterfowl. The water temperature for scalding ducks is normally 135 or 145°F (57–63°C), and for geese 145 to 155°F (63–68°C). The length of the subscald is 1½ to 3 minutes.

Semiscald

With the semiscald the bird is sloshed up and down in water at a temperature of 125 to 130°F (52–54°C). Generally, a water temperature of 128 to 130°F (53–54°C) for 30 seconds gives satisfactory results. The temperature and time vary with the age of the birds. Older birds require a higher temperature and a longer time.

Semiscalding of poultry softens and spreads the fat beneath the surface of the skin, improving the appearance of the dressed bird. This method is generally used for young birds such as broilers, fryers, roasters, and capons. Overscalding is caused by too high a temperature for too long.

Killing, Scalding, and Feather Removal

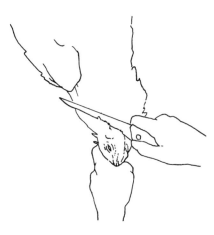

1. Make cut at least 2" (5.1 cm) long at the base of the skull

2. Scald: Follow time and temperature recommendations carefully.

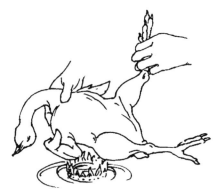

3. Use a rubbing action with the thumbs while picking.

4. Singe over a flame or with a blowtorch.

Wax Picking

One other technique may be used in connection with the semiscald or the subscald, and this is known as *wax picking*. After the birds are scalded, they are dipped in hot wax, then removed from the wax and dipped in cold water to solidify the wax. The bird is coated with wax and the feathers and pinfeathers become embedded in it, so that when the wax is stripped from the carcass, the feathers and pinfeathers are easily pulled out. This method is commonly used in dressing waterfowl. It is not often used when dressing chickens or turkeys.

A modification of the wax-picking method is sometimes used for dressing waterfowl. Paraffin is added to the scald water to form a thick floating layer. The bird is held in the scalding vat for the required length of time and then removed slowly, so that a layer of paraffin adheres to the feathers. The paraffin is given an opportunity to solidify and is then stripped off, carrying with it many of the feathers and pinfeathers that are normally hard to pull.

Waterfowl feathers tend to resist thorough wetting during the scalding process. A little detergent in the scald water will help solve this problem and make the job of picking somewhat easier.

Pinning and Singeing

Pinfeathers, the tiny immature feathers, are best removed under a slow stream of cold tap water. Use slight pressure and a rubbing motion. Those that are difficult to get can be removed by using a pinning knife or a dull knife. Applying pressure, squeeze out the pinfeathers. The most difficult may have to be pulled.

Chickens and turkeys usually have a few hairlike feathers left following the plucking operation. These hairs can be removed by singeing with an open flame. Singeing equipment is available commercially for large operations, but for small units, a small gas torch or a gas range works well. The birds are easily singed by rotating the defeathered bird in the flame.

Eviscerating

After the carcasses are picked and singed, wash them in clean, cool water. As soon as they are washed, they are ready to be eviscerated. Some prefer to cool the poultry before eviscerating and cutting it up, because it is somewhat easier and cleaner to do after it is cooled. Others eviscerate and then cool by placing the birds in ice water or cool water that is constantly replenished. There are many methods of eviscerating poultry. The following methods are not the only ones but they are satisfactory. (See Evisceration, Step-by-Step on pages 255–258.) Neatness and cleanliness are essential for a satisfactory job.

The tools needed for eviscerating poultry are a sharp, stiff-bladed knife, a hook if the leg tendons are to be pulled, and a solid block or bench upon which to work. All equipment and work surfaces should be clean. A piece of heavy parchment paper or meat paper may be laid on the surface and changed as necessary.

Eviscerating Roasting Chickens, Turkeys, and Waterfowl

Roasters, turkeys, and capons are usually stuffed and roasted, so the evisceration procedure is the same for all.

Shanks and Feet

It is best when eviscerating turkeys to remove the tendons from the drumsticks before removing the shanks and feet. By cutting the skin along the shank, the tendons extending through the back of the leg may be exposed and twisted out with a hook or a special tendon puller, if one is available. Many small processors no longer remove tendons.

The shanks and feet should be cut off straight through or slightly below the hock joint, leaving a small flap of the skin on the back of the joint. This will help prevent the flesh on the drumstick from drawing up and exposing the bone during the roasting process. The oil sac, or preen gland, on the back near the tail should be cut out, as it sometimes gives a peculiar flavor to the meat. This is removed with a wedge-shaped cut.

Trachea, Esophagus, Crop, and Neck

To remove the trachea, esophagus, crop, and neck, cut the head off and slit the skin down the back of the neck to just between the shoulder blades. Separate the skin from the neck, and then from the esophagus and trachea. Follow the esophagus to the crop and remove by cutting below the crop. The loose skin then serves as a flap that folds over the front opening and permits stuffing the bird without sewing. Cut off the neck as close to the shoulders as possible. Pruning shears or tree-limb loppers are handy for this purpose. The flap of skin is then folded back between the shoulders and locked in place by folding the tips of the wings over it.

Viscera

Loosen the vent by cutting around it. Do this carefully to avoid cutting into the intestine. Remove the viscera through a short horizontal cut approximately 1½ to 2 inches (3.8–5.1 cm) below the cut made around the vent. The horizontal cut should be about 3 inches (7.6 cm) long. Break the lungs, liver, and heart attachments carefully by inserting your fingers through the front opening. Loosen the intestines from the rear opening by working your fingers around them and breaking the tissues that hold them in the body cavity. Remove the viscera through the rear opening in one mass by inserting two fingers through the rear opening and hooking them over the gizzard, cupping your hand and using a gentle pull and slight twisting motion.

Remove the gonads (ovaries and testes). They are attached to the backbone, and can be removed quite easily by hand. Surgical capons should not have gonads unless they are "slips."

The lungs are attached to the ribs on either side of the backbone; they are pink and spongy in appearance. These can be removed by using your index finger to break the tissues attaching the lungs to the ribs. Merely insert your finger between the ribs and scrape the lungs loose.

Washing

After all the organs have been removed, wash the inside with a hose or under a faucet. Also wash the outside, removing all adhering dirt, loose skin, pinfeathers, blood, and singed hairs. Immediately place the birds into a tank of cold water. This can be running water or water and ice. The birds must be chilled to a carcass temperature of at least 40°F (4°C). After removing them from the chill tank, hang the birds to drain the water from the body cavity, then package and put them into the refrigerator.

Evisceration, Step-by-Step

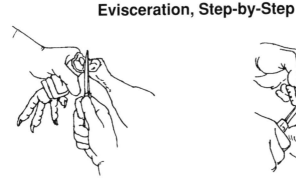

1. *Use boning knife to remove feet at hock joint.*

2. *Cut deeply into tail and remove oil gland.*

3. *Cut off head before the first neck vertebra.*

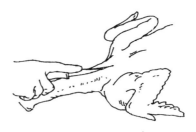

4. *Split neck skin from shoulders to Atlas joint.*

5. *Pull out trachea, esophagus, and crop.*

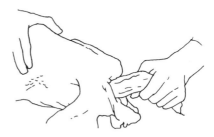

6. *Cut around neck bone, double neck, and twist off.*

(continues on page 256)

Evisceration, Step-by-Step (continued)

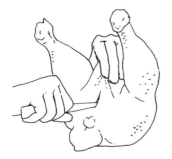

7. *Midline cut: Pull up abdominal skin, then cut through skin and down toward tail.*

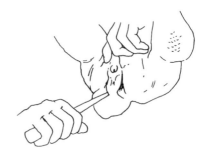

8. *Cut around vent, avoiding cut into intestines.*

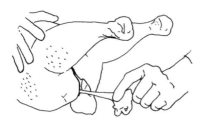

9. *Remove vent and large intestine, and discard.*

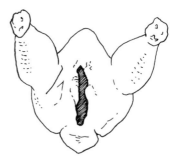

10. *The completed midline cut for broilers and small roasters.*

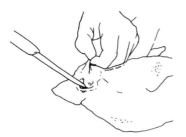

11. *Bar cut: Cut halfway around vent, then . . .*

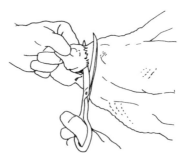

12. *Using fingers as a guide, cut with shears to free vent. Avoid large intestine.*

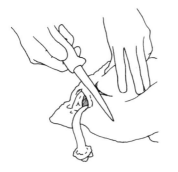

13. *Pinch skin up and make 3" (7.6 cm) horizontal cut.*

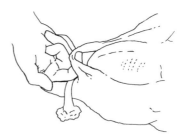

14. *This leaves a "bar" of skin 1½–2" (3.8–5.1 cm) wide.*

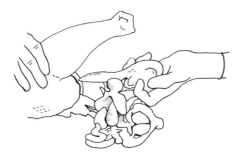

15. *Pull out entrails and save giblets and liver.*

16. *Cut through yellow lining of gizzard, and separate.*

17. *Peel and discard yellow lining and gizzard contents.*

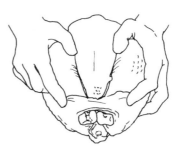

18. *Tuck tail, then legs, under bar strap.*

(continues on page 258)

Evisceration, Step-by-Step (continued)

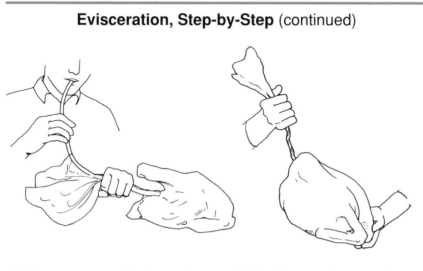

19. *To air-truss, place bird in plastic bag and draw air out of bag with a hose until bird is trussed.*

20. *Twist mouth of bag four times, double, and secure.*

Trussing

A properly trussed bird makes for a nicer appearance. Trussing also conserves the juices and flavors during the roasting process. One method is to place the bird on its back and draw a string across the shoulders and over the wings. Then cross over the drumsticks, bring it under the back, draw tightly, and tie at the base of the tail.

Another method of trussing is to place the hock joints under the strip or bar of skin between the vent opening and the cut from which the viscera were removed. The neck flap is drawn back between the shoulders and the wing tips are folded over the shoulders and serve to hold the skin.

The same procedures are used for killing and eviscerating waterfowl.

Cleaning the Giblets

Remove the gallbladder (the green sac attached to the liver) without breaking it. If the gallbladder is broken while removing the viscera or cleaning the liver, it will likely give a bitter, unpleasant taste to any part with which it comes into contact, as well as cause a green discoloration.

If the gizzard is cool and care is used, it can be cleaned without breaking the inner lining. Cut carefully through the thick muscle until a light streak is observed (do not cut into the inner sac or the inner lining). You may then pull the gizzard muscle apart with the thumbs and remove the unbroken sac, its contents intact.

Eviscerating Broilers and Fryers

The method used for eviscerating broilers and fryers is similar, except that fryers are usually cut into more pieces. Both can be eviscerated the same as roasters, but the usual method is to split and cut up the fryers.

The first operation is to cut off the legs at the hock joint. Remove the oil sac at the base of the tail and cut off the neck. Place the bird on its side, its neck toward you and its back toward your cutting hand. Split the bird with a single cut down the backbone or, if preferred, make two cuts along either side of the backbone and strip out the backbone and neck. The cuts along the backbone should be made with a stiff-bladed knife, just deep enough to sever the ribs. Cut out and remove the vent and the viscera.

To remove the breastbone, simply make a small cut in the cartilage toward the front of the breastbone. By applying a downward pressure on both sides of the bone with your thumbs and snapping it, you can pull out the breastbone. Broilers are usually left whole or in halves, and fryers are quartered.

Eviscerating Fowl

Fowl or old laying hens can be eviscerated similarly to roasters and turkeys. However, they are too tough for barbecuing, roasting, or frying and are usually cut up for stewing or fricassee. One method of cutting up fowl is as follows:

1. Remove the head, neck, crop, esophagus, and trachea as described under roasting chickens, turkeys, and capons.

2. Remove the feet, oil sac, wings, and legs.

3. Make a slight cut on both sides of the body to the rear and below the breast, cutting to the backbone.

4. Break the fowl apart by bending back the breast.

5. Cut away the breast, back and neck, and liver, leaving the viscera in one mass. The breast may be split into halves, if desired.

6. Separate the drumsticks and thighs by cutting through the second joint, the one above the hock joint.

7. Clean the giblets as described earlier.

Chilling and Packaging

It is important that the body heat be removed from the birds as soon as possible after killing. If the cooling is done slowly, bacteria can develop, causing spoilage and undesirable flavors.

If birds are to be air-dried, the air temperature should be 30 to 35°F (1–2°C) to cool the birds properly. The time required to cool the carcasses depends on the size of the birds and the temperature of the air. It may vary from 6½ hours for a 3-pound (1.4 kg) fowl to 10 hours for a 7-pound (3.2 kg) male bird.

Water cooling can be used when air temperatures are above 35°F (2°C). If the birds have been dressed in scalding temperatures that are too high, or for too long a time, the air-dried birds may have a blotchy, discolored appearance. When the hard scald is used, water cooling is the preferred method of cooling the carcasses. Dressed birds should be cooled in pails or tanks of ice water or cold running water. The important factor is maintaining a constant temperature of 34 to 40°F (1–4°C). To cool the birds to an internal temperature of 36 to 40°F (2–4°C) requires 5 to 10 hours in the water, depending again upon the size of the carcass.

For cut-up poultry, the birds may be cut immediately after dressing, thrown into cold water, and cooled in a much shorter time. It should again be noted that it is much easier and cleaner to eviscerate or cut up poultry after it has been cooled.

Cool and age the birds for approximately 8 to 10 hours. If eaten or frozen immediately after dressing, the carcasses will be tougher than if aged for a period of time.

Remove the carcasses from the running water or ice water and hang them up to dry for 10 to 30 minutes before packaging. Every effort should be made to get all of the water out of the body cavity before putting the bird into the bag.

Wrap the giblets in a square sheet of wax paper or a sandwich-sized zip-seal plastic bag. Giblets can be stuffed into the body cavity. They should be wrapped well, so the carcass will not be affected if they spoil.

There are two types of bags available for poultry — one is the common plastic bag and the other is the shrinkable bag (Cryovac) that adheres closely to the bird after shrinking in boiling water. It makes a much more attractive package. It also helps reduce the amount of water loss during the freezing process. Good plastic bags will also do a satisfactory job of maintaining quality in frozen-dressed poultry. The bags should be highly impermeable to moisture to prevent dehydration in freezer storage, which causes toughness.

Thoroughly truss the birds, and then insert them front end first into the plastic bag. The bag should be excessively large. After the bird is in the bag, remove the excess air by applying a vacuum cleaner or by inserting a flexible hose into the top of the bag and applying a vacuum. Merely keep the bag snug around the hose or vacuum cleaner and suck the air out of the bag. Twist the bag several times and secure it with a wire tie or rubber band.

Fresh-dressed, ready-to-cook poultry has a shelf life of approximately 5 days. It should be refrigerated at a temperature of 29 to 34°F (-2–1°C). If it is to be frozen, freeze by the third day after it is dressed and chilled. Chill poultry to below 40°F (4°C) before placing it in the freezer.

Weight losses from live to dressed state vary with the type, age, and size of bird.

Processing Yields — Poultry

TYPE	APPROXIMATE PERCENT YIELD LIVE TO EVISCERATED
Broilers and fryers	75
Roasters	76
Fowl — leghorn type	68
Fowl — heavy type	70
Turkeys — heavy roasters	80
Turkeys — light roasters	78
Ducks	70
Geese	68–73

Source: New England Poultry Management and Business Analysis Manual, *Bulletin 566 (Revised).*

State and Federal Grading and Inspection

Processors of poultry and eggs sold off the farm are subject to the Poultry Products Inspection Act and the Egg Products Inspection Act. There are exemptions for small producers. For information on grading and inspection programs and how they affect you, contact your state department of agriculture.

FROM A FEW PULLETS TO A FLOCK OF 300

After he got a promotion and a raise a few years ago, Bill and his wife Annette moved out of the city to a rural area. The property they purchased had an old chicken coop, so they cleaned up part of it and purchased a few Rhode Island Red pullets, just to supply the family with fresh eggs. Soon they were awash with eggs! They were new to raising poultry and didn't realize that twenty hens produce more than a dozen eggs each day, much more than they would ever need for their family of five.

Bill and Annette and their three children liked working with the chickens, so each child was given responsibility for a specific job each day, such as collecting eggs, filling feeders, and cleaning waterers. The eggs started to pile up, so Bill and Annette began giving away the extra eggs to neighbors and friends at work. Soon they were getting requests for their fresh eggs and a few of their "customers" asked if the eggs were organic or free-range eggs, because they tasted so good and fresh.

Bill, being a businessman, realized that there might be a market for locally produced, fresh eggs from floor-reared chickens. He checked out the local markets, talked with the managers, and found a small demand for his type of eggs. The family decided that it would be a good part-time business, and they all started to learn more about small-scale commercial chicken farming. They contacted the local Extension Poultry Specialist and learned about feeding and managing a flock of floor birds and where to purchase bulk feed and a larger flock of started pullets. Since they already had the land, the coop, and a small customer base, they decided to increase their flock to 300 birds. They picked up a small egg washer and purchased cartons and packed nest-run eggs. Soon they were collecting more than 20 dozen eggs a day and selling them for a good profit at a couple of local markets and directly to the public at the "farm door." They call their business Country Family Eggs and the profits are put into a college fund for the children.

THIRTEEN

FLOCK HEALTH

Disease is not normally much of a problem with small flocks, but there are several diseases that can possibly affect them. It was emphasized earlier that an ounce of prevention is worth a pound of cure. If you purchase stock from a good clean source, follow a sound sanitation program, establish a good biosecurity program, use a good feeding program, and provide a comfortable house with ample dry litter and plenty of fresh air, you will have gone a long way toward keeping your flock healthy.

Recommendations for a sound disease-control program on poultry farms should include the following biosecurity measures:

- Buy chicks or started pullets from disease-free sources.
- Brood chicks away from older birds.
- Do not mix birds from another farm with birds in your flocks.
- Keep visitors away from your poultry buildings; do not permit them to enter pens.
- Screen out wild birds, rodents, and predators from the buildings.
- Clean and sanitize buildings between flocks.
- Obtain a reliable diagnosis before starting treatment for a disease problem.
- Never permit contaminated equipment from other poultry farms in your buildings.
- Use a sound vaccination program.
- Properly dispose of dead birds.

You must realize that losses will occur in any flock. Commercial flock owners expect a monthly mortality of approximately 0.75 percent.

However, a disease in the flock is usually manifested by a drop in feed and water consumption, in egg production, fertility, and hatchability, and sick or dead birds. Birds may look dull or listless. They may show specific signs such as bloody droppings, coughing or sneezing, rattled breathing, paralysis, and other signs that distinguish them from healthy birds. When it is apparent that a disease is present in the flock, seek the advice of a trained poultry diagnostician. It is not advisable to use drugs or antibiotics indiscriminately. Sometimes this will do more harm than good and the only result may be a waste of money.

For the diagnosticians in your area, consult the list at the back of this text. When you submit birds to a state diagnostic laboratory, submit a sample of the flock problem. The sample should include two or more sick birds and freshly dead birds, if any. Take care to preserve dead specimens by cooling to prevent decomposition. (It is not recommended to freeze dead birds, as this may cause cell rupture and make diagnosis more difficult.) An early diagnosis and fast treatment are always recommended for a quick solution to poultry disease problems.

Many of the diseases described here are common to both chickens and turkeys. To avoid duplication, these will be covered in the one section. Those diseases that are peculiar to or more important in waterfowl will be covered in the section on diseases of waterfowl. Diseases of game birds are discussed in chapter 11.

Diseases of Poultry

There are many poultry diseases, but only the more common ones will be discussed here. The diseases will not be discussed in great detail, because this text is intended mainly for small-flock situations. There are many excellent books and pamphlets on poultry diseases if more in-depth information is desired.

Aspergillosis, or Brooder Pneumonia

Aspergillosis, brooder pneumonia, is primarily a disease of young birds, though older birds sometimes have it. Symptoms include dumpy-acting birds, rapid breathing, gasping, and possibly inflamed eyes.

The disease is caused by a fungus, *Aspergillus fumigatus* or another of the *Aspergillus* species. Fungal spores in the air are inhaled by the birds, touching off an infection. Young birds may become infected in contaminated incubators or hatchers, or they may be exposed in the brooder house, where they inhale the fungal spores from musty litter or feed. Older birds are sometimes contaminated by musty pen conditions or feed.

On postmortem, yellow or gray nodular lesions may appear in the lungs, trachea, bronchi, and air sacs. With systemic aspergillosis, lesions are found in the liver, intestine, and brain. There is no known treatment. Further spread of the infection may be prevented by culling the sick birds from the flock and cleaning and disinfecting the pens with a 1 percent solution of copper sulfate or nystatin. Equipment should also be cleaned and disinfected. Moldy or musty litter or feed must be carefully removed to help prevent further spread of the disease.

Avian Influenza

Avian influenza (AI) is a highly infectious viral disease of wild birds and poultry characterized by respiratory symptoms, depression, and reduced feed and water intake. It is caused by orthomyxovirus type A viruses. What makes influenza such a problem is that thirteen H (hemagglutinin) and nine N (neuramidase) types have been identified. The combination of H_xN_x subtypes can quickly mutate from a mild to a more severe form. For example, H_5N_2 was the subtype responsible for the 1983 and 1986 outbreaks in the northeastern United States. The subtypes of H_5 and H_7 appear to be the most deadly of the subtypes encountered thus far.

The viruses have a worldwide distribution, mostly attributed to migratory birds. Avian influenza is readily spread by air, feces, humans, flies, insects, infected litter, dead birds, and poultry manure. It has been isolated in live-bird markets, auctions, and imported pet birds.

The economic losses due to outbreaks make this an especially problematic disease, because once isolated the only recourse is depopulation of the flock. Outbreaks in 1998, 1999, and 2000 in Mexico, Hong Kong, and Italy have illustrated the persistence of this problem.

Two forms of AI have been described. In the mild form, symptoms such as listlessness, respiratory signs such as labored breathing, diarrhea, and low mortality are seen. In the acute form, facial swelling, cyanosis, and dehydration with respiratory distress are seen. Dark red/white spots (cyanosis/ischemia) develop on the legs and combs of chickens. Morbidity is 100 percent and mortality can range from very low to near 100 percent. Mortality increases in an exponential fashion — 10 to 50 times that of the preceding day, peaking at the sixth or seventh day of illness.

There is no treatment for AI at this time. In Mexico, an AI vaccination against the H$_5$ subtype seen in the outbreak was used. Some turkey producers used AI vaccines, but no vaccine for chickens has been approved in the United States. Prevention is the only recourse, and this includes diligent biosecurity measures, as previously described (see page 263).

Blue Comb

The cause of blue comb is not known. It affects both chickens and turkeys, and it can strike birds at any age but usually affects chickens and pullets early, at less than 35 weeks of age, in the laying cycle. Mortality can be heavy. When an outbreak occurs in laying chickens or laying turkeys, egg production is severely affected.

Characteristic symptoms include a loss of appetite, darkening of the head, and a watery diarrhea. The birds are listless and reduce feed and water consumption.

Antibiotics given in the drinking water usually result in a dramatic recovery when treatment is started early. Tetracycline, neomycin, streptomycin, penicillin, and bacitracin have all been used to treat blue comb disease.

Bumblefoot

Bumblefoot is the swelling of the foot involving the foot pad and the area around the base of the toes. It is thought to be caused by injuries resulting either from bruising or a penetration of wire or thorns or other sharp objects with which the feet come in contact. Organisms such as *Escherichia coli*, *Streptococcus*, and *Staphylococcus* have been isolated from the pus material found in the swollen area.

Equipment should be designed so that the birds do not get foot punctures from wire or nails. A deep, dry litter is helpful in preventing this condition, especially in those areas in front of high roosts or nests that require birds to jump to the floor.

Remove birds with severely swollen feet or very lame birds from the flock. Some birds may recover if you remove the scab over the swollen area and squeeze out the core. Then wash the foot with an antiseptic. Application of antibiotic ointments over the open cut helps reduce reinfection.

Chronic Respiratory Disease

Chronic respiratory disease (CRD) is an infectious and contagious disease affecting poultry, game birds, pigeons, and passerine birds of all ages. It is prevalent in all areas of the United States and is caused by a pleuropneumonia-like organism: *Mycoplasma gallisepticum* (MG).

Chronic respiratory disease affects the whole respiratory system, especially the air sacs. It is not a killer but can cause considerable morbidity, especially if treatment is not undertaken soon and secondary infections are not avoided. Both feed efficiency and growth rate are affected in growing flocks. Outbreaks in laying flocks cause lower feed efficiency, and peak egg production and rate of production for the laying cycle will likely be below normal.

In young birds, the signs are rattling, sneezing, and sniffling; in older birds, the condition may go unnoticed. When it becomes complicated, there may be a nasal discharge or a foamy exudate from the eyes. There may be difficult breathing, lack of appetite, and a drop in egg production in producing birds. Young birds may be somewhat stunted and unthrifty.

Postmortem findings include cloudy air sacs with cheesy exudates and mucus in the nasal passages, trachea, and lungs.

Transmission occurs through the hatching egg, through the air, and via infected equipment, feed bags, and other means.

Several antibiotics are used to treat the disease and may be administered in the drinking water, the feed, or by injection. These currently include tetracycline, tylosin, and spectinomycin.

Stress appears to play a key role in triggering the disease. Thus, good environmental conditions and good management are important.

Most breeder flocks in the United States are now MG negative. It is now possible to obtain MG-free stock, and the use of MG-free birds is recommended. Where several age groups are kept on farms and some flocks are MG positive, it is advisable to vaccinate young growing birds to prevent outbreaks of the disease during the laying period, with resulting drops in egg production. If you vaccinate your birds or have a flock that is MG-positive, you should inform your customers.

Coccidiosis

Coccidiosis is primarily a problem in chickens and turkeys, though on rare occasions it may be found in flocks of ducks and geese. It is a common disease of poultry and is caused by a protozoan parasite, coccidium. Birds expose themselves to the disease by picking up sporulated oocysts in fecal material and litter. It should be assumed that all flocks grown on litter are exposed to the disease; birds grown on wire are not exposed to droppings and normally don't become exposed to coccidiosis.

Coccidia are host specific; that is, those that affect chickens do not affect turkeys. Nine species of coccidia affect chickens. Of these nine, three cause most of the problems. Different species affect various parts of the digestive tract. Seven species are known to infect turkeys, but only four of them commonly cause serious problems.

Ruffled feathers, unthriftiness, head drawn back into the shoulders, an appearance of being chilled, and a diarrhea that may be bloody with some forms are all symptoms of coccidiosis. If permitted to go unchecked, considerable mortality, morbidity, and poor flock results may occur.

On postmortem, lesions and hemorrhages may be seen in the various parts of the intestine, depending upon the species involved.

The disease may be prevented by feeding coccidiostats in the growing diet to permit the birds to build an immunity, or completely controlled by feeding a preventive level of coccidiostats. Coccivac, a vaccine containing sporulated oocysts, may be used as a spray on the feed instead of adding a coccidiostat.

Treatment involves the use of sulfonamides, amprolium, or other coccidiostats as prescribed by a diagnostician. Sulfa compounds should be used to treat laying birds only when absolutely necessary, since they severely affect egg production and can cause residues in eggs and tissues for 5 days or more.

Erysipelas

Erysipelas is mostly a disease of turkeys, caused by the bacterium *Erysipelothrix rhusiopathiae*. Erysipelas means "red skin." It may affect ducks and, in rare cases, chickens. Swine, sheep, and humans are also susceptible.

Symptoms are swollen snoods, bluish purple areas on the skin, and congestion of the liver and spleen. Birds may become listless, have swollen joints, and exhibit a yellowish green diarrhea. It is primarily a disease of toms because of injuries obtained during fighting. The erysipelas organism readily enters through skin breaks.

Erysipelas is considered to be a soilborne disease, and contaminated premises are assumed to be the primary source of infection. Sucking and biting insects, especially the mosquito, have been blamed for the spread of erysipelas. Infection enters through bites or breaks in the skin, such as result from pecking or fighting. The disease responds to procaine penicillin. Lincomycin and erythromycin are also effective. Control requires good management and sanitation.

Fowl Cholera

Fowl cholera is caused by a bacterium, *Pasteurella multocida*. It is a highly infectious disease of all domestic birds, including chickens, turkeys, pheasants, and ducks.

The birds become sick rapidly and may die suddenly without showing external symptoms. They may appear listless, feverish, drink excessive amounts of water, and show a diarrhea.

Postmortem findings include red spots or hemorrhages on the surface of the heart, lungs, or intestines, or in the fatty tissues. The birds may have swollen livers (a cooked appearance) with white spots. Treatment with sulfonamides such as sulfaquinoxaline, sulfamethazine, sulfadimethoxine, and others are currently recommended. Sulfaquinoxaline in the feed at the 0.33 percent level for 14 days is considered to be one of the best treatments. Antibiotics are sometimes injected at high levels.

Sanitation in the poultry house, rodent-predator control, range rotation, and proper disposal of dead birds help prevent cholera outbreaks. In problem areas, birds may need to be vaccinated with a bacterin. Thoroughly clean and disinfect buildings and equipment following a cholera outbreak. It is advisable to vaccinate new birds coming onto farms that have had a history of cholera infection.

Fowl Pox

Fowl pox is found in many areas of the country. It is caused by a DNA avian-pox virus. There are six closely related viruses or strains — fowl pox, pigeon pox, canary pox, quail, psittacine, and ratite. It is spread by air, by contact with infested birds, or by mosquitoes. There are two forms of fowl pox: the dry or skin type and the wet or throat type. The same organism causes both. The disease affects most birds — chickens and turkeys, pheasants, quail, ducks, psittacine birds, and ratites — of all ages. Pigeon pox can infect pigeons, chickens, turkeys, ducks, and geese. Canary pox infects canaries, chickens, sparrows, and possibly other species. Fowl pox has been isolated from crows in some parts of the United States.

Birds with fowl pox have poor appetites and look sick. The wet pox causes difficult breathing, a nasal or eye discharge, and yellowish, soft cankers on the mouth and tongue. With the dry pox, small grayish white lumps on the legs, vent, face, comb, and wattles develop. These eventually turn dark brown and become scabs.

On postmortem, cankers may appear in the membranes of the mouth, throat, and trachea. There may be occasional lung involvement or cloudy air sacs.

There is no treatment for the disease itself, though an antibiotic may help reduce the stress of the disease. The only means of control is to administer a vaccine. This is recommended only in those areas where fowl pox is a problem.

Fowl Typhoid

Fowl typhoid is caused by a bacterium, *Salmonella gallinarum*. It affects chickens, turkeys, ducks, pheasants, peafowl, guinea fowl, and other species of birds. The disease may be present wherever poultry is grown; however, it has been all but eliminated in the United States. Most outbreaks occur in exhibition or exotic fowl.

Affected birds may look ruffled, droopy, and unthrifty. Other symptoms may include pale combs and wattles, loss of appetite, increased thirst, and a yellowish green diarrhea.

On postmortem, the liver may have a mahogany color. The spleen may be enlarged and there may be pinpoint necrosis in the liver and other organs. There may be pinpoint hemorrhages in the fat and muscle tissue and evidence of enteritis.

Prevention, treatment, and control are much the same as for pullorum disease, except that there is a vaccine available that is useful in controlling mortality. The blood test used to detect pullorum disease will also detect fowl typhoid in carrier birds.

Histomoniasis or Blackhead

Blackhead is caused by a protozoan parasite. It is infrequently found in chickens but is a common disease of turkeys of all ages. Since it can affect both chickens and turkeys, and chickens may act as an intermediate host for the blackhead-causing organism, it is commonly felt to be risky to keep chickens and turkeys on the same farm. The term *blackhead* is misleading, because this symptom may or may not be present.

Cecal-worm (*Heterakis gallinae*) eggs can harbor the organism over long periods of time, and when picked up by the turkeys, they infect the intestines and liver. Mortality may reach 50 percent if treatment is not started and the infection checked immediately. Symptoms include increased thirst, decreased appetite, watery sulfur-colored droppings, maybe drowsiness, weakness, and dry, ruffled feathers. The head may become cyanotic; hence, the name *blackhead*. Inflammation of the intestine and ulcers on the liver may be seen upon autopsy. The incidence of the disease and its severity depend upon the management and sanitation programs used. Management factors such as sanitation of the brooding facilities, rotation of the range areas, control of cecal and earthworms, and segregation of young birds from old birds help prevent outbreaks. Segregation of turkeys from chicken flocks is helpful in preventing problems with blackhead.

There are currently no approved drugs for treatment or prevention.

Infectious Bronchitis

Infectious bronchitis is another viral disease of poultry. It spreads very rapidly, hits birds of all ages, and is found in chickens. It may cause up to 50 percent mortality in young chicks and 5 to 10 percent in adult birds. It affects the reproductive organs of young birds, thus affecting future performance.

Symptoms include sneezing and coughing. Egg production drops to nearly zero in laying birds and the eggs become misshapen, soft-shelled, chalky or porous, and light in color in the case of brown eggs. Interior egg quality is also affected. Some misshapen eggs may occur after clinical symptoms are gone.

Postmortem findings may be inflamed nasal passages with cheesy plugs in the lower trachea and bronchi. There is no known treatment for infectious bronchitis. Treatment with antibiotics for 3 to 5 days is usually recommended to ward off secondary infections.

Control of infectious bronchitis is accomplished through a vaccination program. Vaccines may be administered by mass methods or individually by nasal or ocular drops. Mass methods of vaccination include drinking water, dust, and spray. Follow the manufacturer's recommendations when using any of these products. Bronchitis vaccine is frequently given in combination with Newcastle vaccine.

Infectious Coryza

Infectious coryza is a respiratory disease of chickens, pheasants, and guinea fowl caused by the bacterium *Hemophilus paragallinarum*. It spreads rapidly from bird to bird and is thought to be triggered by stress. Mortality may be high but is usually low. Symptoms include an involvement of the sinuses and nasal passages with a nasal discharge, sneezing, and swelling of the face. Egg production drops in laying flocks. Postmortem findings usually include congested nasal passages and fluid-filled tissues of the face and wattles.

Old birds from flocks that have had coryza should be treated as carriers and not mixed in with young birds. They should be disposed of or at least kept isolated from flocks that have not had the disease. Treatment with erythromycin is usually beneficial. Some of the sulfonamides are effective but should not be used in layers. Bacterins are available for use in many states.

Prevention is best achieved with an all-in, all-out flock-management program. Most outbreaks occur as a result of mixing flocks.

Infectious Sinusitis

Infectious sinusitis is a contagious disease of turkeys (upper and lower forms), chickens, game birds, pigeons, and passerine birds of all ages, caused by *Mycoplasma gallisepticum* (MG), the same organism that causes chronic respiratory disease (CRD) in chickens (see page 267).

It is transmitted through the eggs from carrier hens; however, since most commercial breeding flocks are MG-free, the disease is most frequently introduced by bringing in infected replacement birds. As in the case of CRD, stress is thought to lower the poults' resistance to the disease.

Symptoms of the upper form for turkeys are watery eyes and nostrils, and a swollen area just below the eyes; for the lower form, airsacculitis with yellow exudates in the air sacs.

In chickens, there may be no outward symptoms or one may see a sticky nasal discharge, airsacculitis, coughing, difficult breathing, foamy secretion in the eyes, and swollen sinuses, accompanied by a drop in feed consumption and loss of body weight. Air-sac involvement may be in evidence on postmortem.

Antibiotics and antibiotic vitamin mixtures will help. Erythromycin, tylosin, spectinomycin, and lincomycin all exhibit antimycoplasma activity. Each, especially tylosin, gives good results in the feed, by water, or by injection.

Infectious Synovitis

Synovitis is an infectious disease of chickens and turkeys caused by *Mycoplasma synoviae* (MS). At first it was identified as a cause of infections of the joints, but more recently it has been identified with a respiratory disease that can affect birds of all ages. It is found in all areas of the United States and is a serious problem, particularly in broiler flocks.

Symptoms include lameness, hesitancy to move, swollen joints and footpads, weight loss, and breast blisters. Some flocks show signs of the respiratory problem. Dying birds may show a greenish diarrhea.

Upon postmortem examination, swelling of the joints and a yellow exudate — especially in the hock, wing, and foot joints — are in evidence. Internally, the birds may show signs of dehydration and enlarged livers and spleens. Those showing the respiratory involvement are not easy to spot. Postmortem examination may show the air sacs filled with liquid exudates.

The most common means of transmission is through infected breeders. Poor sanitation and management practices also contribute. In most primary breeding flocks of chickens, the infection has been eliminated by serological testing. Treatment is prescribed for market egg flocks, but breeder flocks with infectious synovitis should probably be disposed of — it can be passed on to their offspring. Antibiotics yield some results. These should be given by injection or in the drinking water; some prefer to use both simultaneously for best results. Tylosin, erythromycin, spectinomycin, lincomycin, and chlortetracycline are effective.

Laryngotracheitis

Laryngotracheitis is a highly contagious herpes viral disease of chickens and pheasants that spreads rapidly. It may be airborne or spread by contact or carrier birds. It affects all ages of chickens and pheasants. Its effect on birds younger than 4 weeks is mild but may be severe in adult birds. Mortality in some cases may be as high as 60 percent. Laryngotracheitis causes egg production to drop but has little or no effect on egg quality.

Clinical signs develop 6 to 12 days after exposure. The one first noticed may be watery eyes. Other signs of the disease include coughing, sneezing, and shaking of the head to dislodge exudate plugs in the trachea; and as the disease progresses, the birds find it difficult to breathe. The neck is outstretched when the birds inhale, and this may be accompanied by a cawing sound. A bloody mucus may be expelled when the birds cough and sneeze.

On postmortem examination, the throat, trachea, and larynx are inflamed and swollen. There is frequently a bloody mucus or yellowish false membrane in the trachea. Severe inflammation of the larynx and trachea are in evidence and occasional cheesy plugs are found in the trachea. No treatment for laryngotracheitis is effective. In those areas where the disease is a problem, follow a vaccination program.

After an outbreak of laryngotracheitis, some birds remain carriers. These birds are a possible source of infection for nonimmune birds.

Lymphoid Leukosis

This disease has caused serious economic losses to the poultry industry for many years. It is caused by a closely related family of RNA leukoviruses referred to as lympho/sarcoma group, commonly called avian leukosis viruses. The disease is passed from hen to chick via the egg, so the incidence is higher in offspring of infected breeders. The virus has a long incubation period, so signs are not noticeable until birds are 16 weeks or older, and, thus, most of the signs of lymphoid leukosis occur in older birds or laying flocks. Mortality from this disease is seldom a problem in the young. Infected birds begin to develop tumors of the internal organs, particularly the liver, kidney, and spleen, and slowly lose weight and eventually die. Acute death losses are not common with lymphoid leukosis but occur a few at a time over a long period.

There is no known treatment for the disease; however, new technologies such as polymerase chain reaction (PCR) and DNA fingerprinting have made it possible to detect breeders that carry the disease.

Marek's Disease

Marek's disease is a disease of chickens caused by a herpesvirus. It takes many forms but is usually characterized by tumor formations in the muscles, nerves, skin, and intestines. Birds older than 2 weeks may be affected. It can cause losses of 30 percent or more in pullet flocks up to the time of housing.

In the neural form of the disease, tumors in nerves, particularly the sciatic nerve of the leg, result in lameness and paralysis. In the ocular form, birds show "gray eye," or irregularly shaped pupils, and blindness. In the visceral form, one finds tumors of the liver, kidney, spleen, gonads, pancreas, proventriculus, lungs, muscles, and skin. Symptoms are lameness; droopy wing(s); incoordination; paleness; weak, labored breathing; and gray iris.

There is no treatment for Marek's disease, and the best control measure is vaccination of the baby chicks. This is usually done at the hatchery. Many hatcheries now vaccinate the eggs just prior to hatching, thus reducing the handling of newly hatched chicks.

Newcastle Disease

Newcastle disease is widespread. It is acute, highly contagious, and found in chickens, turkeys, ducks, and geese, and all other birds of all ages. Humans and other mammals are also susceptible. It is a respiratory disease caused by a virus. It causes high mortality in young flocks, and production in laying birds frequently drops to zero.

Newcastle spreads rapidly through the flock, causing gasping, coughing, and hoarse chirping. Water consumption increases and a loss of appetite occurs. Infected birds tend to huddle and exhibit signs of partial or complete paralysis of the legs and wings. They may hold their heads between their legs or on their backs with their necks twisted.

The disease is transmitted in many ways. It can be tracked in by people or brought in with chickens from another premise, or it can come from dirty equipment or feed bags, or by wild birds that gain entrance to the pen.

Postmortem findings may include congestion and hemorrhages in the gizzard, intestine, and proventriculus. The air sacs may be cloudy and exudates may be found in the lungs, trachea, and air sacs.

There is no effective treatment, though antibiotics are normally given to reduce the chance of secondary infections. Vaccination is recommended in most areas of the country and can be administered individually or on a mass basis. Recommendations vary with the area and the type of bird. Newcastle disease vaccine can be administered intranasally, ocularly, or by wing web. On a mass basis, it can be given alone or in combination with infectious bronchitis vaccine — in the drinking water or in the form of a dust or spray. Consult the manufacturer's recommendations for use of the product. Follow the vaccination program recommended for your area and the type of bird to be grown.

Omphalitis, Navel Ill, Mushy Chick Disease

Omphalitis is caused by bacterial infection of the navel occurring when it doesn't close properly after hatching. It is found in young chicks and turkey poults, and may be caused by poor incubator or hatchery sanitation, excessive humidity in the hatcher, and chilling or overheating of the chicks.

Birds with omphalitis are weak and unthrifty and tend to huddle together. The abdomen may be enlarged and feel soft and mushy. The navel is infected, and the area around it may be a bluish black. Mortality may be high for the first 4 to 5 days.

There is no treatment for the disease. Most of the affected chicks die the first few days. No medication is needed for the survivors. Prevention is the only solution.

Paratyphoid

Paratyphoid is an infectious disease of chickens, turkeys, ducks, and other birds and animals. It is caused by one or more of the salmonella bacteria. There are several types, but *Salmonella typhimurium* is one of the most common in poultry and it accounts for more than half of the typhoid outbreaks in poultry flocks. Transmission may come from the hen through the egg and to the chick. The organism is also found in fecal material of infected birds.

The disease is primarily one of young birds, but older birds may also be affected. Acute outbreaks are seen in birds 7 to 21 days of age, with peak mortality from 7 to 14 days. In young birds, mortality may run as high as 50 percent. Some birds may die without showing symptoms, while others show signs of weakness, loss of appetite, diarrhea, and pasted vents. Birds may appear chilled and huddle together for warmth. In older birds, there is a loss of weight, weakness, loss of egg production, and diarrhea.

Postmortem of young birds may show unabsorbed yolk sacs, small white areas on the liver, inflammation of the intestinal tract, congestion of the lungs, and enlarged livers. Older birds usually show no lesions, though a few may show white areas on the liver.

Sulfamerazine, nitrofurans (where legal; the FDA withdrew it from the U.S. market in 1991), and some of the antibiotics may reduce losses, prevent secondary invaders, and increase appetite.

Control is through sanitation and isolation of the flock from sources of infection such as wild birds, birds from other flocks, and contaminated feed and equipment.

Pullorum

Pullorum is an infectious disease of chickens, turkeys, and some other species caused by *Salmonella pullorum*. It is found all over the world. Pullorum is highly fatal to birds under 2 weeks of age.

Mortality begins at 5 to 7 days. The birds appear droopy, huddle together, act chilled, and may show diarrhea and pasting of the vent. Pullorum is sometimes called white diarrhea.

Salmonella pullorum spreads from the hen to the chick through the egg. Chicks' down is a vector in incubators and hatchers.

Postmortem findings in infected chicks include dead tissue in the heart, liver, lungs, and other organs, and an unabsorbed yolk sac. In adult birds, there are discolored and misshapen ova; the heart muscle may be enlarged and show grayish white nodules. The liver may also be enlarged, yellowish green, and coated with exudate. Several types of blood tests help establish a positive diagnosis for pullorum.

Sulfamethazine and sulfamerazine are effective in reducing mortality. Nitrofurans, where legal, may also be effective in reducing mortality, but will not cure the disease. Medication may hinder diagnosis.

Flocks that have recovered from pullorum should not be kept for replacements or breeding purposes unless they have been blood-tested and found to be free from the disease.

Diseases of Waterfowl

Waterfowl raised in small flocks seldom have disease problems. Where waterfowl are kept in large numbers and concentrated in relatively small areas, disease may be more prevalent. Some breeds of waterfowl exhibit more resistance than others. For example, the Muscovy duck appears to be more resistant to diseases common in the Pekin and the Runner.

Only the more common diseases of waterfowl will be considered in this section. It should again be emphasized that when disease problems occur, it is best to get a reliable diagnosis from a poultry diagnostician.

Amyloidosis

This is one of the common diseases of adult ducks and geese. Amyloidosis is also called wooden-liver disease and can be readily recognized by the hardness of the liver. Mortality may reach as high as 10 percent in some flocks. The cause of the disease is not known and there is no known cure. Postmortem may show large accumulations of fluids within the body cavity.

Aspergillosis, or Brooder Pneumonia

This disease was discussed earlier under chicken and turkey diseases. The causes and control measures are the same for waterfowl. (See page 264.)

Botulism

Botulism occurs in both young and adult waterfowl and in other types of poultry. It is caused by the bacterium known as *Clostridium botulinum*. This organism grows in decaying plant and animal material. Birds feeding on material containing the neurotoxins produced by the bacteria lose control of their neck muscles. For this reason, it is sometimes referred to as "limber neck." Infected waterfowl may drown if swimming water is available to them. Maintaining dry, clean facilities and removing decaying vegetation or spoiled feed and removing and properly disposing of dead birds will usually prevent this disease. A mixture of Epsom salts, at the rate of 1 pound (0.4 kg) per 5 gallons (19 L) of water, has been reported as an effective treatment.

Coccidiosis

As mentioned earlier, coccidiosis is an infrequent problem with ducks. The type of protozoan causing the disease in ducks is different from those causing the problem in chickens and turkeys, but control and treatment are the same as for other species of birds.

Duck Virus Enteritis, or Duck Plague

Duck plague affects ducks, geese, swans, and other aquatic birds. It is caused by a virus that is transmitted by contact. It commonly occurs where water is available for swimming and has affected ducks in commercial flocks in the eastern United States.

Signs of the disease are watery diarrhea, a nasal discharge, and general droopiness. The symptoms develop 3 to 7 days after exposure. Death frequently results in 3 or 4 days. Generalized hemorrhages in the body organs may be seen on postmortem.

There is no satisfactory treatment. Only strict sanitation and rearing ducks in pens with access to drinking water will reduce and help control the disease.

Exposure of domestic ducks to duck virus enteritis from migratory waterfowl can be avoided by keeping the susceptible ducks in houses or wire enclosures.

Fowl Cholera

Fowl cholera, which also attacks ducks and geese, has been discussed earlier under chicken and turkey diseases. Control and treatment of fowl cholera in waterfowl are the same as for other birds. (See page 269.)

Keel Disease

Keel disease occurs in young ducklings the first few days after hatching. Affected birds appear thin and dehydrated.

Fumigation of the hatching eggs and a thorough washing and fumigation of the hatcher between hatches are recommended to reduce the number of bacteria to which the young ducklings are exposed. Clean, warm brooding facilities and good feed and fresh water will also help control the disease.

New Duck Syndrome, or Infectious Serositis

Caused by a bacterium, *Pasteurella anatipestifer*, new duck syndrome is one of the most serious diseases affecting ducklings and may also be seen in young turkeys. The bacteria invade the bloodstream to produce a generalized septicemia. The symptoms resemble those of chronic respiratory disease of chickens. The first signs of the disease are coughing, sneezing, and ocular and nasal discharge — followed by a greenish diarrhea, loss of balance, tremors, coma, and death. Losses as high as 75 percent have been recorded. Death often is due to water starvation rather than to the primary infection. Antibiotics and sulfa drugs have been used with some success.

Viral Hepatitis

Viral hepatitis outbreaks can cause 80 to 90 percent mortality in flocks of young ducklings. It is a highly contagious disease and strikes ducklings from 1 to 5 weeks of age.

Vaccination of the female breeding stock will prevent this disease. Antibodies produced by the vaccinated laying ducks are passed on through the eggs to the young ducklings. It is recommended that ducklings be brooded and reared in strict isolation, especially for the first 5 weeks, to help outbreaks of the disease.

Parasites of Poultry

Although there are many parasites of poultry, relatively few of them are of major importance. Some of the parasites live inside the bird and others live on the outside.

Internal Parasites

Poultry are commonly susceptible to many internal parasites, including nematodes, acanthocephalans, cestodes, trematodes, and protozoa. Some can cause setbacks in weight gain, loss of egg production, and even death, if infestation is severe. A few of the intestinal parasites harbor other disease organisms harmful to chickens and turkeys.

Nematodes

The more commonly seen internal parasites are discussed here.

Large Roundworm. A light infestation of roundworm (*Ascaridia galli*) probably does little damage to the bird. When worms become numerous, however, the bird can become unthrifty and feed conversion can suffer. The worms themselves seldom cause mortality, but when present with other diseases, death may result. The worms are species specific, with little or no cross-infection from one class of poultry to another.

The large roundworm is 1½ to 3 inches (3.8–7.6 cm) long. It is found in the middle of the small intestine and causes setbacks in weight gain and a loss of egg production in adult birds. The piperazine compounds are used to treat roundworms and can be given in the water, in the feed, or in capsule form. Levamisole hydrochloride (Tramisol) can also be used in the water; meldane (Coumaphos) and Hygromycin B may be used in the feed. The addition of extra vitamin A in the diet for 5 to 7 days will speed recovery of the intestinal mucosa from damage by ascarid migration.

Birds infect themselves with the roundworm by picking up the eggs from feces, the litter, or other infested materials. It is therefore desirable to keep the litter as clean as possible to prevent the birds from picking up the eggs. Birds on wire floors or in cages do not have worms.

Capillaria Worm. The capillaria worm is a small worm ½ to 1½ inches (1.3–3.8 cm) in length. It inhabits the upper part of the intestine, including the esophagus, crop, and occasionally the ceca. It can cause inflammation of the intestines, diarrhea, and weakness, and in laying birds, lowered egg production. It is difficult to see capillaria worms with the naked eye, so sometimes their presence is overlooked on autopsy. Treatment rations are often supplemented with additional vitamin A, because capillaria-worm infestations prevent efficient utilization of the vitamin in the bird. Meldane or Hygromycin B added to the diet has been reported to be helpful in some cases.

Capillaria has the same direct life cycle as the roundworm. The embryonated egg is picked up from the litter by the bird. The earthworm is also an intermediate host. Sanitary conditions are important for the control of capillaria worms. It is difficult to get rid of them where dirt floors are used in poultry buildings.

Cecal Worm (*Heterakis gallinarum*). The life cycle of the cecal worm is similar to that of the large roundworm. These worms are found in the ceca instead of the small intestine. Their length is ½ to ¾ inches (1.3–1.9 cm). By itself, it probably has little economic significance in chickens, but its eggs harbor the protozoan organisms that cause blackhead disease of turkeys, which is why the two birds should not be raised together. In severe infestations, cecal worms may contribute to a somewhat unthrifty condition, weakness, and emaciation. It is rather hard to treat, because most of the intestinal contents bypass the ceca during the process of digestion. Phenothiazine and Hygromycin B have been used successfully as treatments.

Gapeworm. The gapeworm, *Syngamus trachea*, attacks the bronchi, trachea, and lungs. It can cause pneumonia, gasping for breath, or even suffocation. High mortality may occur in young birds, especially turkeys and pheasants.

The use of thiabendazole, where legal, in the feed for 2 weeks is effective in eliminating gapeworms. Levamisole hydrochlorine has been reported as being an effective treatment when given in the drinking water for 3 days.

Gizzard Worm. The gizzard worm is reddish in color, about ½ to 1 inch (1.3–2.5 cm) in length. It is found under the lining of the gizzard, and it impairs digestion due to the damage to the gizzard. No treatment is known, but the disease may be controlled by avoiding exposure of the birds to such intermediate hosts as grasshoppers, beetles, weevils, and other insects.

Cestodes: Tapeworm. There are several species of tapeworms varying in size from microscopic to 6 to 7 inches (15.2–17.8 cm) long. They are flat, white, and segmented. They inhabit the small intestine and cause a loss of weight and a loss of egg production in adult laying birds. Treatment recommended for tapeworms is dibutyltin dilaurate (Butynorate; Tinostat) or one of the triple dewormers on the market, such as Valbazen (Pfizer, Groton, CT).

The effective control of internal parasites depends primarily upon a program of cleanliness and sanitation. Parasitic eggs can remain viable in the soil for more than a year. This makes it important for producers to practice rotation of poultry runs or yards and, if possible, to keep them clean and under cultivation to prevent the birds from picking up the parasites.

External Parasites

All types of poultry suffer from various lice, mites, fleas, and bugs. Some can affect growth and egg production, while others cause symptoms so severe that they may lead to lameness, blindness, and even death.

Lice

Lice are chewing or biting insects that cause birds considerable grief. With severe infestations, egg production, growth, and feed efficiency can suffer. They cause an irritation of the skin with a scab formation. Sucking lice do not infest domestic fowl. The mature louse is wingless, flat-bodied, and six-legged, with double claws and a round head.

Lice spend their entire lives on the birds; they will die within a few hours if they leave their host. The average female louse lays from 50 to 300 eggs. The eggs are laid on the feathers, where they are held with a gluelike substance. They hatch in a few days to 2 weeks. Lice live on the scales of the skin and feathers. They do not usually stay on waterfowl that have access to swimming ponds.

There are several types of lice that attack poultry, including the body louse, shaft louse, wing louse, large chicken louse, and brown chicken louse. The body louse is one of the most common lice of poultry and usually affects older birds. The lice and its nits (eggs) are seen on the fluff, the breast, under the wings, and on the back. The head louse attacks the feathers of the head and seems to be most harmful to young birds.

Materials that may currently be used to treat lice are carbaryl (Sevin), malathion, and coumaphos (Co-Ral). Use them according to the directions on the label and examine birds frequently for signs of a reinfestation.

Northern Fowl Mite

Northern fowl mites are a reddish dark brown in color. They infect chickens, turkeys, and game birds, especially pheasants, and are found around the vent, the tail, and the breast of the birds. These mites live on the birds at all times and may be seen on eggs in the nest when there is a severe infestation. They attack the feathers, suck blood, and cause anemia, loss of weight, and loss of egg production. Materials recommended for treatment of the northern fowl mite include carbaryl, coumaphos, and permethrin. Use according to directions.

Red Mite, or Chicken Mite

Red mites are not found on the birds during the daytime — they feed during the night. They may be seen on the underside of roosts, in cracks on the wall, or in seams of the nest during the daylight hours. Other signs are salt-and-pepper-like trails under roost perches or clumps of manure on the furniture. Red mites are bloodsuckers. They cause irritation, loss of weight, loss of egg production, and anemia. Laying birds may refuse to use infested nests. Use carbaryl, coumaphos, or malathion according to the manufacturer's directions.

Depluming Mite

The depluming mite burrows into the skin and causes an irritation at the base of the feather. In an attempt to relieve the itching, the birds pull out feathers until they are nearly naked in cases of severe infestation.

To control the depluming mite, use a dip containing 2 ounces (56.7 g) of sulfur and 1 ounce (28.3 g) of soap per gallon (3.8 L) of water. Dip the birds to wet the feathers to skin level. Dip only on warm days. Repeat the treatment in 3 or 4 weeks, if necessary.

Tropical Chicken Flea

This flea clusters on the comb, wattles, and earlobes and around the eyes of the bird. It causes irritation to the membranes of the eye and may cause eventual blindness and death. It also causes a loss of weight and lowered egg production. Use carbaryl, coumaphos, or malathion according to the manufacturer's directions.

Fowl Tick, or Blue Bug

The fowl tick, also known as the blue bug, is a soft, flat bug. Mature fowl ticks are ¼ to ½ inch (6.3–12.7 mm) in length, leathery, egg-shaped, eight-legged, and thin-bodied, and are reddish to blue-black in color. The larvae are six-legged and ⅒ inch (2.5 mm) long and dark blue or purplish in color. The adults are nocturnal feeders. They cause a loss of appetite and of body weight, as well as lowered production. Loss of blood may result in anemia. The ticks may also carry diseases: Heavy infestation may lead to so-called tick paralysis. Carbaryl, coumaphos, and malathion are effective materials for the control of ticks, or blue bugs.

Bed Bug
The bed bug is a brown or yellow-red bug found under roosts and in cracks and crevices. They are wingless and about ¼ inch (6.3 mm) long. Their oval body is quite flat until engorged with blood. The bugs are nocturnal feeders; they do not remain on the birds during the daylight hours. Bed bugs prefer warm environments and, unless a heated building is provided, are not much of a problem in winter. Birds pull out their own feathers in an effort to relieve the intense itching caused by bed bugs. The treatment for this parasite is the same as for the northern fowl mite.

Scaly Leg Mite
This eight-legged mite is microscopic in size. It burrows into the skin of the legs and under the scales, and may occasionally attack the comb and wattles. It may cause the birds to go lame or be crippled for life. There is evidence of scales and crusts on the legs, which may be swollen. Scaly leg mite is not common where proper sanitation is practiced.

Treatment is on an individual-bird basis. Dip legs into a mixture of 1 part kerosene and 2 parts raw linseed oil, mineral oil, or warm petroleum jelly. Repeat this procedure once or twice at weekly intervals until the problem is solved.

Chiggers
The chigger that attacks poultry is the same red mite that attacks humans. Mature chiggers usually are covered with dense feathered hairs, giving them a velvety appearance. The body is figure-eight-shaped and often bright red and is about ½₅ inch (1 mm) long. Birds infested with chiggers may become droopy and emaciated. Chiggers attach themselves to the birds' skin in clusters under the wings and on the back and neck. They may cause abscesses and inflamed areas. Birds may even die with severe infestations.

One aid for controlling chiggers is to keep the grass and weeds on range or in the yards cut short. You may also dust the range with 1 percent malathion at the rate of 40 pounds (18.1 kg) per acre.

Precautions for Drug and Pesticide Use

To get optimum production results from poultry, it is sometimes necessary to use drugs or pesticides. Use them with caution. First, make certain of the problem and then use the material of choice according to recommendations. *Never* use a material that is not registered for use on poultry.

Neither drugs nor pesticides are intended as substitutes for good preventive management. They work best when used in combination with good sanitation and sound management practices. Early diagnosis and treatment of disease or parasitic problems is recommended.

Agencies of the federal government have set up definite withdrawal periods and tolerance levels for various materials used in poultry. The Food and Drug Administration, for example, has set minimum periods of time when certain drugs must be withdrawn prior to poultry slaughter. These withdrawal periods vary from 1 to 2 days up to several days, and are subject to change from time to time. Maximum tolerances for residues of certain materials are also established. Some insecticides can be used around poultry but not directly on the birds, on the eggs, or in the nests. Some pesticide materials cannot be used within a certain number of days prior to slaughter. Frequency of use is also an important consideration with some materials. Before using any product, be sure that it is legal. Follow the instructions for use exactly, and dispose of any unused materials and containers according to the manufacturer's instructions.

Use Caution

Withdrawal periods and tolerances change and accepted treatment materials also change, so specific precautions for use of various materials will not be addressed here. Drugs and insecticides should be used discriminately, following all the precautions specified on the label for the use of a given material. Improperly used, they can be injurious to humans, animals, and plants.

Store all drugs and pesticides in the original containers in a locked storage area. Keep them out of the reach of children and animals. Avoid prolonged inhalation of sprays or dusts, and wear protective clothing and equipment as recommended. Be safe!

Nutritional Diseases

A number of poultry diseases are caused by nutritional deficiencies or imbalances. With the well-formulated diets on the market today, nutritional problems are uncommon. Therefore, a thorough discussion of nutritional diseases will not be undertaken. To point out the importance of good nutrition, however, a few of the more common nutritional diseases will be mentioned.

Rickets

Rickets is caused by a deficiency or imbalance of calcium, vitamin D_3, or phosphorus. Birds with rickets appear to be weak, have stiff, swollen joints, soft beaks, soft bones of the leg, and enlarged or beaded rib bones. Phosphorus rickets causes soft, rubbery bones.

Perosis, or Slipped Tendon

Perosis is a leg weakness usually caused by a deficiency of manganese in the diet. Choline, niacin, and biotin may also have a role. With this deficiency, most chicks will respond to manganese fortification in the feed. Symptoms are flattening and enlargement of the hocks, followed by slippage and lateral rotation of the Achilles tendon out of the condyles of the hock. When this occurs, the shank and foot remain flexed and extended laterally from the body. There is no treatment for affected birds. In general, poultry require 35 to 50 parts per million of manganese in their feed to prevent this problem. Usually, 3 to 5 days of supplement with a multiple mineral-vitamin supplement in the drinking water will prevent new cases from developing.

Encephalomalacia

Encephalomalacia, or vitamin-E deficiency, is commonly called crazy chick disease because it causes staggering, incoordination, and paralysis of affected chicks. Vitamin-E deficiency causes a part of the brain to degenerate. Positive diagnosis of this disease requires a study of the brain tissues. There is no treatment for affected birds, but supplementation with vitamin E will prevent new cases.

Curled-Toe Paralysis

Curled-toe paralysis is caused by a riboflavin (vitamin B$_2$) deficiency in the feed. Feed conversion, growth rate, and bird livability (mortality rate) are adversely affected. One or both feet may be affected. Additional fortification of the feed with riboflavin will help to prevent further development of this disorder.

Vitamin-A Deficiency

Birds deficient in vitamin A become unthrifty. Exudates may form in the eyes and nostrils and whitish pustules develop in the mouth, esophagus, and crop. Mortality in young poultry can be high if the deficiency is not corrected. Adult birds exhibit reduced production, lowered hatchability, and decreased livability in the offspring. You might also see an increased incidence of blood spots in eggs of birds with vitamin-A deficiency. Yellow corn, legumes, grasses, and fish oils are good sources of vitamin A.

Predators

Rats and mice are a common problem on poultry farms. They like to eat poultry feed. They may destroy feed bags that are hung where they can get at them. They destroy insulation in poultry buildings and are carriers of disease. A few rats or mice will quickly produce a large population if not controlled. Rats are cautious and mice are curious. The Norway rat, the most common on poultry facilities, is an excellent swimmer; the roof rat, found near seacoasts and in the southern United States, is an excellent climber. Rats and mice are opportunistic feeders; although rats tend to eat most of their food in one sitting, mice eat sporadically. Rats need a source of water, while mice can live on the moisture they gain from feed, so keep water spills and pools cleaned up. Usually, if you have mice, you don't have rats, and vice versa.

Rats and mice like hiding places. Given places to hide in or around the chicken house, they will set up housekeeping. They like to inhabit the space between double-sheathed walls, stone foundations, under floors, and in and around junk. Good housekeeping will help eliminate hiding spots and prevent rodent problems.

Trapping can be helpful if the populations of rats and mice are small. Keeping a cat in the area will also help control them. However, one must be aware that cats can also spread diseases, such as salmonella, and should not generally be in contact with the birds in a coop or house. Baiting is the best way of handling large populations.

Anticoagulant baits are the most effective means of eradicating rats and mice. Anticoagulants cause rodents to bleed to death internally after they have consumed the bait for several days.

To be effective, bait stations should be set up near walls, burrows, and paths that the rats and mice tend to follow. These bait stations must be

Baiting Stations

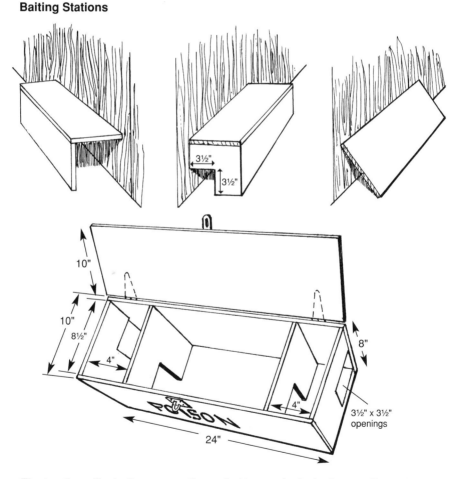

The top three illustrations are wall-type baiting stations; the bottom illustration is a box-type baiting station.

kept full of fresh bait — but no more than the rats or mice will eat in 2 or 3 days. If bait gets old and musty, they will not touch it. Place fresh water near the bait stations to increase consumption of the bait. Choosing the right rodenticide for your particular rodent problem is important.

Multiple-dose anticoagulant baits must be eaten for several days consecutively to kill the rodent. Multiple-dose nonanticoagulant baits are made with vitamin-D complex and are usually not harmful to other animals. Single-dose anticoagulant rodenticides require only a single feeding to kill the rodent. These are particularly toxic to dogs, so take care to remove dead rodents from the area. Acute single-dose rodenticides are useful for a quick knockdown of a rodent population that has gotten out of control. It is best to rotate baits on a regular basis, so you don't develop a bait-shy rodent population.

Homemade Bait Station

A simple bait station can be made from 1½-inch (2.8 cm) PVC pipe. Cut three 1-foot-long (30.5 cm) sections of pipe and connect with a T joint. Set the pipe on the ground as an upside-down T and attach it to a wall of your building, inside or out. Place bait into the top leg of the T and cap it off. Corner bait stations can be made using a 90-degree T with the legs following either wall and the upper section in the corner.

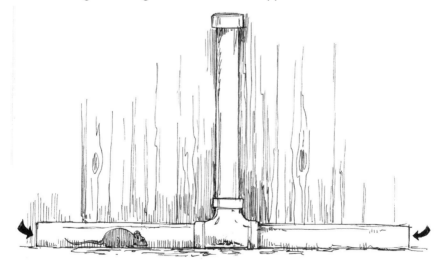

Types of Rodenticides

Drug Type/Name	Dose	Comment(s)
ANTICOAGULANT		
Brodifacoum	Single	Most potent rodenticide currently available for commensal rodents; good to use with warfarin-resistant rats. With a single feeding, rats die within 4 or 5 days.
Bromadiolone	Single	Less potent than brodifacoum, but better than chlorophacinone or diphacinone; rats die within 5 days after a single feeding.
Chlorophacinone	Multiple	Rodents must feed several times but not necessarily on consecutive days. For optimal effect, several feedings should occur within a 10-day period.
Coumafuryl	Multiple	For an effective kill, the rodents need several feedings.
Diphacinone	Multiple	Works the same as chlorophacinone, and either formulation can be used.
Isovaleryl-indandione	Multiple	It is used primarily in tracking powders for controlling rodents that are not eating bait.
Pindone	Multiple	One of the early bait formulations; slightly less effective than warfarin, but still kills rats.
Warfarin	Multiple	First marketed anticoagulant; best known but used less today because of development of more potent rodenticides, like brodifacoum and bromadiolone.
DIPHENALYAMINE		
Bromethalin	Single, acute*	A single-dose rodenticide that causes central nervous system depression and paralysis leading to death in 2–4 days
DIPHOSPHIDE		
Zinc phosphide	Single, acute*	A dark gray powder that produces phosphine when it comes in contact with the dilute acids in the stomach of animals after ingestion, killing the animal overnight.
STEROL (VITAMIN-D COMPLEX)		
Cholecalciferol	Multiple	This is a single- or multiple-dose rodenticide that causes mobilization of calcium from bones into the blood, causing death from hypercalcemia in 3 or 4 days. Very safe with other animals, as rats are more sensitive to vitamin-D toxicity at very low doses.

*A single feeding kills the rodent in a matter of hours.

Important Note

Ideally, grain should be stored in rat- and mouseproof bins or containers. Covered galvanized garbage cans with plastic-bag liners are excellent for storing feed for small flocks and can discourage rodents.

Other Predators

Although rats and mice are the biggest nuisance to small flocks, you should be aware of other possible problems. At times, dogs and cats may raid the poultry yard or house. Cats can be a problem if they gain access to a pen of young chicks. Occasionally, dogs get into the poultry house or poultry yard and manage to kill young or adult birds. Usually, the real loss comes from piling of the birds and subsequent suffocation due to fright.

Coyotes, foxes, raccoons, skunks, weasels, and mink may frequent areas where poultry is kept. The work of a fox can be identified by the signs of hair on the fence, or a hole through which it has crawled. He usually kills several birds, and frequently leaves them behind, partially buried. When the raccoon comes to visit, it tends to eat the crops out of the birds, and some heads may be missing. Given the opportunity, it returns every fourth or fifth night to get more.

Owls and hawks also like chicken. If one or two birds are killed every night and the heads and necks are missing, an owl may be the problem.

When birds are killed occasionally and are badly beaten, a dog problem is indicated. If several birds are killed on an irregular basis, it may also be the work of a mink or a weasel, in which case there will be small bites around the neck and head.

Loss Prevention

Confinement rearing has prevented many losses from predators. Birds ranged or yarded should be closed in a shelter at night. If range shelters are used, screen the space beneath the shelter. An electric fence a few inches above the ground discourages some predators. A strong metal fence buried 18 to 24 inches (45.7–61.0 cm) underground, with a 10- to 12-inch (25.4–30.5 cm) J hooking away from the enclosure, backfilled with rocks or stone, also works well.

Dead-Bird Disposal

Disposal of dead birds in a sanitary manner reduces the spread of disease as well as fly and odor problems. Dead birds can be placed in disposal pits, incinerated, buried, or composted. Those flock owners who are more fortunate may have a pickup service available.

Disposal Pit

A properly constructed disposal pit is an efficient and practical way to dispose of dead birds. They are relatively inexpensive to build and require little maintenance. Construct the pit in well-drained soil at least 100 feet (30.5 m) away from wells or springs and where drainage is away from other water supplies in the neighborhood. You must check your local zoning and health regulations and obtain necessary permits before installing a disposal pit or burying birds.

The disposal pit is capable of handling the mortality from flocks of about 1,000 birds for 3 years. For smaller flocks, the pit size can be scaled down somewhat.

Deep burial is a practical and satisfactory method of dead-bird disposal for small flocks. It must be done properly, however. If buried too shallowly, the carcasses may be dug up by dogs and other animals.

Disposal Pit for Dead Birds

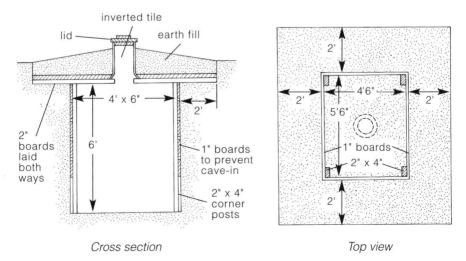

Cross section   Top view

Incineration

A good incinerator is an excellent means of dead-bird disposal. There may be objectionable odors to contend with, especially if there are neighbors nearby. Modern incinerators are quite efficient and most have afterburners, which tend to reduce the odors. The main disadvantages are the necessary investment and the high operating costs, particularly for small-flock use.

Composting

A composter can be built for any size poultry farm and is based on the number of birds to be composted. A general rule is 1 pound (0.5 kg) of dead chicken per cubic foot (0.03 cu m) of primary compost space per day. A roof over the compost area is required for proper composting of dead birds. An impervious floor, usually concrete, is also required, if all-weather composting is to be accomplished. A simple minicomposter using old wooden shipping pallets for the walls works well.

A simple formula, or "recipe," for composting consists of a simple mixture of poultry litter, straw, and dead chickens in a ratio of 1 part chicken, 1½ parts litter, and ⅒ part straw.

Moisture content of the mixture should be about 40 to 50 percent, as evidenced by a squeezed handful of compost mix leaving wetness on the palm of your hand without forming drops of water. Contact your local Extension Poultry Specialist for plans.

THE FAMILY PET

For a high school project, Stacey wanted to see if red contact lenses really would reduce cannibalism in chickens, as claimed by the manufacturer. She contacted the Extension Poultry Specialist and together they set up an experiment with twenty New Hampshire Red hens. Stacey actually purchased twenty-four pullets, to make sure she would have twenty for the project. The company in Massachusetts that produced the lenses donated them for the test, and it was decided that ten pullets would be fitted and ten would not. The four other birds would be extras, just in case any got sick or died.

Well, the lenses fell out of the birds' eyes, since they had been meant for smaller Leghorn-type chickens, so Stacey decided to rear half under red lights and half under white. Red lights made the birds less cannibalistic. However, during this time, one of the birds got its leg caught in something in the yard and twisted it and cut it up. It became infected and Stacey's mother took the bird to the veterinarian, who amputated the leg at the hock. This bird was only 22 weeks old at the time, and Stacey and her mom kept the bird in the house during its recovery. It soon learned to hop along on one leg and became the family pet, almost never leaving Mom's side. If the family dog or cat got too near to Mom, the chicken, called Elsie, would screech and attack it. Elsie considered the house her castle and kept a good guard on it. She became litter-box trained and even went to the door and pecked at it when she wanted to go outside.

Long after the 23 other hens were gone, and Stacey was a member of the poultry club at college, Elsie, the one-legged chicken, was still the family pet, sitting on Mom's lap when she watched television and standing by her side in the kitchen, waiting to catch a scrap of food while Mom prepared meals. A friend of the family thought the one-legged poultry pet so unusual, she called a newspaper reporter, who wrote an article, complete with pictures, about Elsie and the family. Elsie lived to be 7 years old, but memories of that one-legged chicken will last a lifetime.

REFERENCES

Aho, William A., and Daniel W. Talmadge. *Incubation and Embryology of the Chick*. Storrs, CT: Cooperative Extension Service, College of Agriculture and Natural Resources, University of Connecticut, 1973.

Ash, William J. *Caponizing Chickens*. Leaflet 490, USDA, 1961.

———. *Culling Hens*. Farmers Bulletin 2216, Agricultural Research Service, USDA, 1966.

———. *Duck Rations*. Extension Stencil 25, Ithaca, NY: Department of Poultry Science, Cornell University, Ithaca, NY.

———. *Egg Grading Manual*. Agricultural Handbook 75. Consumer and Marketing Service, USDA, 1977.

———. *Embryology and Biology of Chickens*. Burlington, VT: The Vermont Extension Service, University of Vermont,

———. *Farm Poultry Management*. Farmers Bulletin 2197. Agricultural Research Service, USDA, 1977.

———. *Raising Ducks*, Farmers Bulletin 2215, USDA, 1966.

Brook, Munro. 4-H In-School Embryology Program, In-Vitro Demonstration. Burlington, VT: University of Vermont Extension Service.

Hauver, W. E., and L. Kilpatrick. *Poultry Grading Manual*. Agricultural Handbook 31. USDA, revised 1989.

Hunter, J. M. and John C. Scholes. *Profitable Duck Management*, 7th ed. Cayuga, NY: The Beacon Milling Co., 1943.

Jordan, H. C., and L. D. Schwartz. *Home Processing of Poultry*. University Park, PA: College of Agriculture Extension Service, Pennsylvania State University, 1974.

Marsden, Stanley J. *Turkey Production*. Agricultural Handbook 393. Agricultural Research Service, USDA, 1971.

Mercia, L. S. *Drawing Poultry*. Burlington, VT: 4-H Publication, Vermont Extension Service, University of Vermont,

————. *The Family Laying Flock*. Brieflet 1218. Burlington, VT: Vermont Extension Service, University of Vermont,

————. *Killing, Picking and Cooling Poultry*. Burlington, VT: 4-H Publication, Vermont Extension Service, University of Vermont, .

————. *Lighting Replacements and Layers*. Burlington, VT: Vermont Extension Service, University of Vermont,

Moyer, D. D. *Virginia Turkey Management*. Publication 302. Blacksburg, VA: Extension Division, Virginia Polytechnic Institute, 1969.

Orr, H. L. *Duck and Goose Raising*. Publication 532. Guelph, Ontario, Canada: Ontario Ministry of Agriculture and Food, Department of Animal and Poultry Science, Ontario Agriculture College, 1969.

Ota, Hajime. *From Egg to Chick, A 4-H Manual of Embryology and Incubation*. Northeast Cooperative Publications.

————. *Housing and Equipment for Laying Hens*. Miscellaneous Publication 728. Agricultural Research Service, USDA, 1967.

————. *Poultry Management and Business Analysis Manual for the 80's*. Bulletin 566 (revised). Cooperative Extension Services, Extension Poultrymen in New England, 1980.

————. *Raising Geese*. Farmers Bulletin 2251. Northeastern Region Agricultural Research Service, USDA, 1983.

————. *Raising Livestock on Small Farms*. Farmers Bulletin 2224. USDA, Revised 1972.

Ridlen, S. F., and H. S. Johnson. *4-H Poultry Management Manual — Unit 1*. Urbana, IL: Cooperative Extension Service, University of Illinois.

Talmadge, Dan W. *Brooding and Rearing Baby Chicks*. Cooperative Extension Services of the Northeast, 1973.

Warren, Richard. *4-H Laying Flock Management*. 4-H Circular 79. Cooperative Extension Services in New England, 1966.

APPENDIXES

General Feed Requirements
for Different Types of Birds at Various Ages

TYPE OF BIRD	AGE	TYPE FEED	PERCENT PROTEIN	LB/100 BIRDS WHITE EGG	BROWN EGG
Chickens:					
Pullets	0–8 wk	Starter	20	280	364
	8–12 wk	Grower	17	308	392
	12–22 wk	Grower or Developer	14	1,064	1,289
Layers	22+ wk	Layer	17	23	25
Broilers	0–4 wk	Starter	22	200	
	5–7 wk	Grower	20	500	
Fryers	7–8 wk	Finisher	18	100	
Roasters	0–4 wk	Starter	22	200	
	4–13 wk	Grower	16	1,200	
	13–market	Finisher	14	1,000	
Turkeys:					
Small	0–7 wk	Starter	28	600	
	7–18 wk	Grower	20	3,000	
Large	0–8 wk	Starter	28	1,000	
	8–16 wk	Grower	20	3,000	
	16–24 wk	Finisher	14	3,500	
Ducks:					
White Pekin for market	0–2 wk	Starter	22–24	214	
	2–4 wk	Grower	18–20	510	
	4–8 wk	Finisher	16–17	1,410	
For Breeders					
	0–2 wk	Starter	22–24		
	2 wk– maturity	Grower	18–20	214 *	
Geese:					
White Emden or Chinese	0–3 wk	Starter	20–22	525	
	3–14 wk	Grower	15	5,600	
	14 wk– market (24–30 wk)	Grower	15		

Ducks and particularly geese are good foragers. From approximately 3 weeks of age with good forage available, savings of up to 35% in feed consumption can be realized.

"Natural" Laying Hen Diet

(To be fed as all-mash to medium-sized layers kept in floor pens)

INGREDIENT	AMOUNT (LB)	
	100 LB	PER TON
Yellow corn meal	60.00	1,200
Wheat middlings	15.00	300
Soybean meal (dehulled)	8.00	160
Maine herring meal (65%)	3.75	75
Meat & bone meal (47%)	1.00	20
Skim milk, dried	3.00	60
Alfalfa leaf meal (20%)	2.50	50
Iodized salt	0.40	8
Limestone, grd. (38% Ca)	6.35	127
TOTALS	100.00	2,000

CALCULATED ANALYSIS	NATURAL LAYING HEN DIET (PER LB)	RECOMMENDED (NECC) NUTRITIONISTS (PER LB)
Metabolizable energy cal./lb	1,252	1,292
Protein	16.07%	16.00%
Lysine	0.79%	0.74%
Methionine	0.31%	0.29%
Methionine and cystine	0.55%	0.54%
Fat	3.67%	3.33%
Fiber	3.15%	2.51%
Calcium	2.77%	2.75%
Total phosphorus	0.53%	0.50%
Available phosphorus	0.44%	0.42%

VITAMINS (UNITS OR MG/LB)

Vitamin A activity (U.S.P. units/lb)	5,112.00	5,290.00
Vitamin D (ICU)		1,000.00
Riboflavin (mg)	1.36	1.38
Pantothenic acid (mg)	3.89	4.05
Choline chloride (mg)	411.00	500.00
Niacin (mg)	17.46	16.95

The birds must receive direct sunlight to enable them to synthesize vitamin D. Unfortified cod liver oil can be fed in place of sunlight to supply vitamin D. The amount of cod liver oil would depend upon the potency of the oil — the need is for 1000 (ICU) per pound of feed. (Prepared by Dr. Richard Gerry, Department of Animal & Veterinary Sciences, University of Maine, Orono)

NECC = New England College Conference nutritionists; ICU = International Chick Unit

Poultry Manure Information

Manure Production

Layers: 25 pounds per 100 per day with normal drying.
Four-tenths of a cubic foot per 100 per day.
Moisture content 75–80% as defecated.
Pollution load — 7 to 12 layers equal 1 human.
Weight of a cubic foot of poultry manure at 70% moisture approximately 65 pounds.

Broilers: Manure and litter per 100 birds.
8 weeks equals 400 pounds or 11.2 cubic feet at 25% moisture.

Fertilizer Value of Poultry Manure*

	MOISTURE %	NITROGEN	POUNDS PER TON PHOSPHORUS	POTASH
Fresh Manure	75.0	29	10	8
Stored Manure	63.9	24	13	16
Broiler Litter	18.9	72	25	30
Layer Litter	22.1	50	23	36
Slurry (liquid)	92.0	22	12	7

When nitrogen is worth 25¢ per pound, phosphorus 15¢ per pound, and potash 8¢ per pound, the values per ton of poultry manure are:

Fresh Manure $9.39
Stored Manure   9.23
Broiler Litter   24.15
Layer Litter   14.86

Other elements of plant food contained in poultry manure include calcium, magnesium, copper, manganese, zinc, chlorine, sulfur, and boron.

*Based on information from Univ. of New Hampshire Bulletin 444 — Farm Manure, 1966.

Directory of Poultry Publications

A number of USDA and state Extension publications on poultry production and marketing are available and may be obtained by contacting the local county Extension agent and state or area Poultry Extension Specialists. The addresses of the poultry specialists, by state, are given in appendix A. A partial list of trade publications and books follows.

Trade Publications

Egg Industry, Watt Publishing Co., Mount Morris, IL 61054
Feedstuffs, Miller Publishing Co., Minnetonka, MN 55343
Poultry Digest, Watt Publishing Co., Mount Morris, IL 61054
The Poultry Times, Poultry Egg and News, Inc., 345 Green Street, NW, Gainesville, GA 30501
Turkey World, Watt Publishing Co., Mount Morris, IL 61054
WATT PoultryUSA, Watt Publishing Co., Mount Morris, IL 61054

Books

Austic, Richard, and Malden Nesheim. *Poultry Production* (13th ed). Philadelphia, PA: Lippincott Williams & Wilkins, 1990.

Damerow, Gail. *Storey's Guide to Raising Chickens*. Pownal, VT: Storey Books, 1996.

Florea, J. H. *ABC of Poultry Raising: A Complete Guide for the Beginner or Expert*. New York, NY: Dover, 1997.

Holderread, Dave. *Storey's Guide to Raising Ducks*. Pownal, VT: Storey Books, 2001.

Mercia, Leonard S. *Storey's Guide to Raising Turkeys*. Pownal, VT: Storey Books, 2001.

Moreng, Robert, and John Avens. *Poultry Science and Production*. Prospect Heights, IL: Waveland Perss, 1991.

North, Mack O., and Donald D. Bell. *Commercial Chicken Production Manual* (4th ed). New York, NY: Chapman & Hall, 1990.

E-Mail and Internet Web Sites for Poultry Information

Poultry News is a list-serve for all people interested in poultry — from commercial growers, integrators, Extension specialists, and researchers to backyard poultry fanciers. Questions are answered by people from all around the world. To subscribe, send an e-mail to:

PLTRYNWS@ sdsuvm.sdstate.edu

In the body of the message, write "subscribe."

poultry.gatech.edu
This site has links to almost everything to do with poultry.

GALLUS.TAMU.EDU/Diseases/AMOPD.htm
This is a good site for learning about avian diseases.

www.poultryegg.org
This is the site for the U.S. Poultry and Egg Association.

www.ansc.purdue.edu/poultry
This links to Purdue University, a source of more than 500 publications on poultry for small-flock producers to commercial producers.

The-coop.org/links/links.html#A15
This links to information on subjects including small flocks, poultry shows, breeds, and poultry clubs.

www.animalscience.ucdavis.edu/extension/avian/
This links to a University of California site, also good for information on production, management, and health.

www.the-coop.org/index.html
This links to one of the original poultry sites. Many links to other good sites.

www.afn.org/~poultry/poulint.htm
This site also provides links to many good sites dealing with poultry.

There are hundreds of good poultry Web sites; however, space does not permit a complete listing.

Diagnostic Laboratories by State

*Designates Laboratories Accredited by the American Association of Veterinary Laboratory Diagnosticians

Not all laboratories are full-service avian veterinary laboratories.

Alabama

State Veterinary Diagnostic
  Laboratory
Wire Road
P.O. Box 2209
Auburn, AL 36831-2209
334-844-4987

State Veterinary Diagnostic
  Laboratory
501 Usury Avenue
Boaz, AL 35957
256-593-2995

State Veterinary Diagnostic
  Laboratory
P.O. Box 409
Hanceville, AL 35077
256-352-8036

Alaska

State Federal Laboratory
500 S. Alaska Street
Palmer, AK 99645
907-745-3236
www.state.ak.us/dec/deh

Arizona

*Arizona Veterinary Diagnostic
  Laboratory
University of Arizona
2831 N. Freeway
Tucson, AZ 85705
520-621-2356
www.microvet.arizona.edu

Arkansas

Arkansas Livestock and Poultry
  Commission Laboratory
3559 N. Thompson Street
Springdale, AR 72764
501-751-4869

Arkansas Poultry Federation
P.O. Box 828
Springdale, AR 72764
501-375-8131

California

California Animal Health and
  Food Safety Laboratory Systems
Fresno Branch
2789 S. Orange Avenue
Fresno, CA 93725
559-498-7740

San Bernardino Laboratory
105 W. Central Avenue
P.O. Box 5579
San Bernardino, CA 92412
909-383-4287

Turlock Laboratory
1550 N. Soderquest
Turlock, CA 95380
209-634-5837

Colorado

Colorado State University,
  Branch Diagnostic Laboratory
27847 Road 21
Rocky Ford, CO 81067
719-254-6382

Connecticut

*University of Connecticut
  Department of Pathobiology
61 N. Eagleville Road
Box U-89
Storrs, CT 06269
860-486-3736

Delaware

Poultry and Animal Health
2320 S. Dupont Highway
Dover, DE 19901
800-282-8685

University of Delaware
Lasher Laboratory
Poultry Diagnostic Laboratory
Route 6, Box 48
Georgetown, DE 19947
302-856-1997
www.rec.udel.edu

Department of Animal and Food
  Sciences
University of Delaware
531 S. College Avenue
Newark, DE 19717
302-831-2524

Florida

*Kissimmee Diagnostic Laboratory
Florida Department of Agriculture
2700 N. John Young Parkway
Kissimmee, FL 34741
407-846-5200

Live Oak Diagnostic Laboratory
912 Nobles Ferry Road
Live Oak, FL 32060
904-362-1216

Georgia

Georgia Poultry Laboratory
  Veterinary Poultry Diagnostic
  Laboratory
180 McClure Street, P.O. Box 349
Canton, GA 30114
770-479-2901

Georgia Poultry Laboratory
410 N. Park Avenue
Dalton, GA 30720
706-278-7306

Georgia Poultry Laboratory
150 Tom Freyer Drive
Douglas, GA 31535
912-384-3719

Georgia Poultry Laboratory
4457 Oakwood Road
Oakwood, GA 30566
770-535-5996

Hawaii
Veterinary Laboratory Branch
Hawaii Department of Agriculture
99-941 Halawa Valley Street
Aiea, HI 96701
808-483-7100

Idaho
Division of Animal Industries
2230 Old Penitentiary Road
Boise, ID 83712
208-332-8570

Illinois
*Animal Disease Laboratory
9732 Shattuc Road
Centralia, IL 62801
618-532-6701

*Animal Disease Laboratory
Illinois Department of
  Agriculture
2100 S. Lake Storey Road
P.O. Box 2100X
Galesburg, IL 61402-2100
309-344-2451

*Veterinary Diagnostic Lab
College of Veterinary Medicine
University of Illinois
Urbana, IL 61802
217-333-1620
www.cvm.uiuc.edu

Indiana
Animal Disease Diagnostic
  Laboratory
11367 E. Purdue Farm Road
Dubois, IN 47527
812-678-3401

*Animal Disease Diagnostic
  Laboratory
1175 ADDL
Purdue University
West Lafayette, IN 47907
765-494-7440
www.addl.purdue.edu

Iowa
USDA National Veterinary
  Services Laboratory
P.O. Box 844
Ames, IA 50010
515-239-8200
www.aphis.usda.gov/vs/nvsl/index.
  html

*Veterinary Diagnostic
  Laboratory
College of Veterinary Medicine
Iowa State University
Ames, IA 50011
515-294-1950

Kansas
*Veterinary Diagnostic Laboratory
College of Veterinary Medicine
Kansas State University
Manhattan, KS 66506
785-532-5650
www.vet.ksu.edu

Kentucky
*Murray State University
Breathitt Veterinary Center
715 North Drive, P.O. Box 2000
Hopkinsville, KY 42240
720-886-3959
www.murraystate.edu/cit/bvc

*Livestock Disease Diagnostic
  Center
University of Kentucky
1429 Newtown Pike
Lexington, KY 40511
859-253-0571
www.ca.uky.edu/lddc

Maine
Animal Disease Diagnostic
  Laboratory
University of Maine
332 Hitchner Hall
Orono, ME 04469
207-581-2775/2788

Maryland
Maryland Department of
  Agriculture Animal Health
  Laboratory
8077 Greenmead Drive
College Park, MD 20740
301-935-6074

Maryland Department of
  Agriculture Animal Health
  Laboratory
1840 Rosemont Avenue
Frederick, MD 21702
301-663-9568
www.mda.state.md.us

Maryland Department of
  Agriculture Animal Health
  Laboratory
P.O. Box 376
Oakland, MD 21550
301-334-2185

Maryland Department of Agri-
  culture Animal Health
  Laboratory
Quantico Road, P.O. Box 2599
Salisbury, MD 21801
410-543-6610

Michigan
*Animal Health Diagnostic
  Laboratory
Michigan State University
P.O. Box 30076
Lansing, MI 48909-1315
517-353-0635
www.ahdl.msu.edu

Minnesota
*Veterinary Diagnostic
  Laboratory
College of Veterinary Medicine
University of Minnesota
1333 Gortner Avenue
St. Paul, MN 55108
612-625-8787
www.mvdl.umn.edu

Poultry Testing Laboratory
622 Business Highway 71 NE
P.O. Box 126
Willmar, MN 56201
320-231-5170

Mississippi
Central Laboratory
P.O. Box 1510
Forest, MS 39074
601-469-4421

Mississippi Veterinary Diagnostic
  Laboratory
P.O. Box 4389
Jackson, MS 39296
601-354-6089

Missouri
Northwest Missouri Veterinary
  Diagnostic Laboratory
302 West Grand
Cameron, MO 64429
816-632-6595

*Veterinary Medical Diagnostic
  Laboratory
University of Missouri
P.O. Box 6023
Columbia, MO 65205
573-882-6811
www.cvm.missouri.edu/vmdl

State-Federal Cooperative
  Animal Health Laboratory
216 El Mercado Plaza
Jefferson City, MO 65109
573-751-3460

Springfield Veterinary Diagnostic
  Laboratory
701 N. Miller Avenue
Springfield, MO 65802
417-895-6863

Montana
*Veterinary Diagnostic Laboratory
Department of Livestock
Diagnostic Laboratory Division
P.O. Box 997
Bozeman, MT 59771
406-994-4885

Nebraska
*Lincoln Diagnostic Laboratories
University of Nebraska–Lincoln
Fair Street and East Campus Loop
Lincoln, NE 68583
402-472-1434

Nevada
Nevada Department of
   Agriculture
Animal Disease Laboratory
350 Capitol Hill Avenue
Reno, NV 89502
775-688-1182, ext. 232

New Hampshire
New Hampshire Veterinary
   Diagnostic Lab
University of New Hampshire
Kendall Hall
129 Main Street
Durham, NH 03824
603-862-2726

New Jersey
New Jersey Animal Health
   Diagnostic Lab
John Fitch Plaza
P.O. Box 330
Trenton, NJ 08625
609-292-3965

New Mexico
New Mexico Veterinary
   Diagnostic Services
700 Camino de Salud NE
Albuquerque, NM 87106
505-841-2576

New York
Cornell Duck Research
   Laboratories
Veterinary Diagnostic Lab
192 Old Country Road
Eastport, NY 11941
631-325-0600

*Veterinary Diagnostic
   Laboratory
New York State College of
   Veterinary Medicine
Cornell University
Ithaca, NY 14853
607-253-3900

North Carolina
Animal Disease Diagnostic
   Laboratory
Paradise Road, P.O. Box 38
Edenton, NC 27932
252-482-3146

Hoyle C. Griffin Animal Disease
   Diagnostic Laboratory
401 Quarry Road, P.O. Box 2183
Monroe, NC 28111
704-289-6448
www.agr.state.nc.us/vet

*Rollins Animal Disease
   Diagnostic Laboratory
2101 Blue Ridge Road
Raleigh, NC 27607
919-733-3986
www.ncagr.com/vet/lab.htm

Rose Hill Animal Disease
Diagnostic Laboratory
329 Yellowcut Road, P.O. Box 37
Rose Hill, NC 28458
910-289-2635

North Dakota
*Veterinary Diagnostic
Laboratory
North Dakota State University
Van Es Hall
Fargo, ND 58105
701-231-8307
www.ndsu.nodak.edu/ndsu/
veterinary_science/vetdiag

Ohio
*Animal Disease Diagnostic
Laboratory
8995 E. Main Street
Reynoldsburg, OH 43068
614-728-6220
www.state.oh.us/agr/addl

Oklahoma
State Department of Health
Laboratory
P.O. Box 24106
Oklahoma City, OK 73124
405-271-5070

*Animal Disease Diagnostic
Laboratory
College of Veterinary Medicine
Oklahoma State University
Stillwater, OK 74078
405-744-6623
www.cvm.okstate.edu

Oregon
OSU Veterinary Diagnostic
Laboratory
College of Veterinary Medicine
Oregon State University
P.O. Box 429
Corvallis, OR 97339
541-737-3261
www.vet.orst.edu

Oregon State Department of
Agriculture
Animal Health Laboratory
635 Capital Street NE
Salem, OR 97301
503-986-4686
www.oda.state.or.us

Pennsylvania
*The Pennsylvania Veterinary
Laboratory (PADLS)
2305 N. Cameron Street
Harrisburg, PA 17110
717-772-2852

Diagnostic Laboratory
New Bolton Center
382 West Street Road
Kennett Square, PA 19348
610-444-5800

Animal Diagnostic Laboratory
Orchard Road
The Pennsylvania State University
University Park, PA 16802-1110
814-863-0837

South Dakota

Department of Veterinary
  Science
Animal Disease Research and
  Diagnostic Laboratory
South Dakota State University
Box 2175
Brookings, SD 57007
605-688-5171
www.vetsci.sdstate.edu

Tennessee

University of Tennessee College
  of Veterinary Medicine
  Diagnostic Services
P.O. Box 1071
Knoxville, TN 37901
865-974-7262
www.vet.utk.edu

C. E. Kord Animal Disease
  Laboratory
Melrose Station
P.O. Box 40627
Nashville, TN 37204
615-837-5125
www.state.tn.us/agriculture

Texas

*Texas Veterinary Medical
  Diagnostic Laboratory
Texas A&M University
Drawer 3040
College Station, TX 77841
409-845-9000

Utah

Utah State University
Veterinary Diagnostic Laboratory
North Logan, UT 84322
435-797-1900

USU Provo Veterinary
  Laboratory
2031 S. State Street
Provo, UT 84606
801-373-6383

Virginia

Harrisonburg Laboratory
116 Reservoir Street
Harrisonburg, VA 22801
540-434-3897
www.vdacs.state.va.us

Ivor Laboratory
34591 General Mahone Blvd.
Ivor, VA 23866
757-859-6221
www.vdacs.state.va.us

Lynchburg Laboratory
4832 Tyreeanna Road
Lynchburg, VA 24504
804-947-2518
www.vdacs.state.va.us

Warrenton Laboratory
272 Academy Hill Road
Warrenton, VA 20186
540-347-6385
www.vdacs.state.va.us

Wytheville Laboratory
250 Cassel Road
Wytheville, VA 24382
540-228-5501
www.vdacs.state.va.us

Washington
*Animal Disease Diagnostic
  Laboratory
Washington State University
College Station
P.O. Box 2037
Pullman, WA 99165
509-335-9696
www.vetmed.wsu.edu

Avian Health Laboratory
Washington State University
Puyallup, WA 98371
253-445-4537
www.vetmed.wsu.edu

West Virginia
State Federal Cooperative
  Animal Health Laboratory
1900 Kanawha Blvd. East
Charleston, WV 25305-0172
304-558-2214

Wisconsin
Wisconsin Veterinary Diagnostic
  Laboratory
1521 E. Guy Avenue
Barron, WI 54812
715-637-3151

*Wisconsin Veterinary
  Diagnostic Laboratory
University of Wisconsin
6101 Mineral Point Road
Madison, WI 54705
608-262-5432

Wyoming
*Wyoming State Veterinary
  Laboratory
1174 Snowy Range Road
Laramie, WY 82070
307-742-6638

Canada
* Animal Health Monitoring
  Lab
1767 Angus Campbell Road
Abbotsford, BC V3G 2M3
CANADA
604-556-3135

*University of Guelph
Animal Health Laboratory
P.O. Box 3612
Guelph, ON N1H 6R8
CANADA
519-824-4120

Diagnostic Services
Atlantic Veterinary College
University of Prince Edward
  Island
550 University Avenue
Charlottetown, PE C1E 1Z4
CANADA
902-566-0863

COOPERATIVE EXTENSION SYSTEM OFFICES BY STATE

The Cooperative Extension System is an excellent source of information on poultry and other agricultural enterprises. Extension personnel are located in county or area offices in each state, as well as at agricultural colleges of the state universities. Poultry specialists are located at many state universities, but not all of them. In some large poultry-producing states, specialists may be located in various areas of the state, as well as on state-university campuses. For information, contact the Cooperative Extension System office nearest you.

Alabama
Auburn University
Poultry Science Department
236 Animal Science Building
Auburn, AL 36849-5416
334-844-2600
www.ag.auburn.edu/dept/ph/
  index.html
www.aces.edu

Tuskegee University
Rm. 202 Morrison/Mayberry Hall
Tuskegee, AL 36088
334-724-4441

Alaska
Cooperative Extension Service
College of Rural Alaska
University of Alaska Fairbanks
P.O. Box 756180
Fairbanks, AK 99775-6180
907-474-7246
www.uaf.edu/coop-ext

Arizona
University of Arizona
Forbes Building, Room 301
Tucson, AZ 85721
520-621-7209
www.ag.arizona.edu/extension

Arkansas
University of Arkansas
Poultry Science Department
Fayetteville, AR 72701
501-575-4952
www.uark.edu/depts/posc
www.uaex.edu

University of Arkansas at Pine
  Bluff
1890 Cooperative Extension
  Service
1200 N. University Drive
Pine Bluff, AR 71601
870-543-8529/8534
www.uapb.edu

California
University of California at Davis
Department of Avian Science
Davis, CA 95616-8521
530-752-3519
animalscience.ucdavis.edu/
extension/avian/

University of California
111 Franklin Street, 6th Floor
Oakland, CA 94607-5200
510-987-0060
danr.ucop.edu

Colorado
Colorado State University
1 Administration Building
Fort Collins, CO 80523-4040
970-491-6281
www.colostate.edu/Depts/
CoopExt/

Connecticut
University of Connecticut
Young Building, Room 231
1376 Storrs Road, U-36
Storrs, CT 06269
860-486-1987
www.canr.uconn.edu/ces/index.
html

Delaware
University of Delaware
113 Townsend Hall
Newark, DE 19717
302-831-2501
ag.udel.edu

Florida
University of Florida
Department of Dairy and Poultry
Sciences
P.O. Box 110920
Gainesville, FL 32610
352-392-1981
dps.ufl.edu/

Georgia
University of Georgia
Department of Poultry Science
4 Towers Building
Athens, GA 30602
706-542-1325
www.uga.edu/~poultry/
www.uga.edu/caes/

Hawaii
University of Hawaii
3050 Maile Way, Room 202
Honolulu, HI 96822
808-956-8234
www2.ctahr.hawaii.edu

Idaho
University of Idaho
P.O. Box 442332
Moscow, ID 83844-2332
208-885-6639

Illinois
University of Illinois
214 Mumford Hall
1301 W. Gregory Drive
Urbana, IL 61801
217-333-5900
www.extension.uiuc.edu

Indiana

Purdue University
Department of Animal Science
1151 Smith Hall
West Lafayette, IN 47907-1151
765-494-8011
ag.ansc.purdue.edu/poultry/

Iowa

Iowa State University
218 Beardshear Hall
Ames, IA 50011
515-294-6192
www.exnet.iastate.edu

Kansas

Kansas State University
123 Umberger Hall
Manhattan, KS 66506
785-532-5820
www.oznet.ksu.edu

Kentucky

University of Kentucky
Department of Animal Science
604 W.P. Garrigus Building
Lexington, KY 40546
606-257-7529
www.uky.edu/Agriculture/Animal
  Sciences/Poultry/ukpoultry.htm
www.ca.uky.edu

Louisiana

Louisiana State University
Department of Poultry Science
Ingram Hall
Baton Rouge, LA 70894
504-388-4406
www.coa.lsu.edu
www.lsuagcenter.com

Maine

University of Maine
5741 Libby Hall, Room 102
Orono, ME 04469
207-581-2811
www.umext.maine.edu

Maryland

University of Maryland at
  College Park
Department of Animal and
  Avian Sciences
College Park, MD 20742
301-405-1373
ansc.umd.edu/ansc2000/index.
  html

University of Maryland, Eastern
  Shore
Backbone Road
Princess Anne, MD 21853
410-651-6206

Massachusetts

University of Massachusetts
  Extension Office
Draper Hall
University of Massachusetts
Amherst, MA 01003
413-545-4800
www.umass.edu/emext/

Michigan

Michigan State University
108 Agriculture Hall
East Lansing, MI 48824
517-355-2308

Minnesota

University of Minnesota
Coffey Hall, Room 240
1420 Eckles Avenue
St. Paul, MN 55108
612-624-2703

Mississippi

Alcorn State University
1000 ASU Drive, 479
Lorman, MS 39096
601-877-6128
www.alcorn.edu

Mississippi State University
Department of Poultry Science
P.O. Box 5188
Mississippi State, MS 39762
662-325-3416
www.msstate.edu/dept/poultry/
www.ext.msstate.edu

Missouri

University of Missouri
309 University Hall
Columbia, MO 65211
573-882-7754
outreach.missouri.edu/

Montana

Montana State University
P.O. Box 172040
115 Culbertson Hall
Bozeman, MT 59717-2040
406-994-6647
www.montana.edu

Nebraska

University of Nebraska
211 Agriculture Hall
Lincoln, NE 68583
402-472-2966
ianrhome.unl.edu/extension1.
html

Nevada

University of Nevada at Reno
Cooperative Extension
National Judicial College 118,
Mail Stop 404
Reno, NV 89557-0106
775-784-7070
www.nce.unr.edu

New Hampshire

University of New Hampshire
Cooperative Extension
Taylor Hall
59 College Road
Durham, NH 03824
603-862-1520
ceinfo.unh.edu

New Jersey

Rutgers Cooperative Extension
Rutgers, The University of
New Jersey
88 Lippman Drive
New Brunswick, NJ 08901
732-932-5000
www.rce.rutgers.edu

New Mexico
New Mexico State University
Department 3AE
P.O. Box 30003
Las Cruces, NM 88003
505-646-3016
cahe.nmsu.edu

New York
Cornell University
365 Roberts Hall
Ithaca, NY 14853
607-255-2116
www.cce.cornell.edu

North Carolina
North Carolina A&T State
  University
P.O. Box 21928
Greensboro, NC 27420
336-334-7691
www.ag.ncat.edu

North Carolina State University
Extension Poultry Science
P.O. Box 7603
Raleigh, NC 27695-7608
919-515-2621
www.ces.ncsu.edu/depts/poulsci

North Dakota
North Dakota State University
315 Morrill Hall
P.O. Box 5437
Fargo, ND 58105
701-231-8944
www.ext.nodak.edu

Ohio
Ohio State University Extension
Ohio State University
3 Agricultural Administration
  Building
2120 Fyffe Road, Room 4
Columbus, OH 43210
614-292-4067
ohioline.ag.ohio-state.edu

Oklahoma
Oklahoma State University
Department of Animal Science
101 Animal Science Building
Stillwater, OK 74078-6051
405-744-5398
www.ansi.okstate.edu/
www.dasnr.okstate.edu

Oregon
Oregon State University
Department of Animal Sciences
101 Ballard Extension Hall
Corvallis, OR 97331-6702
541-737-5066
www.orst.edu/Dept/animal-
  sciences/poultext.htm
osu.orst.edu/extension/

Pennsylvania
Pennsylvania State University
Department of Poultry Science
213 William N. Henning
  Building
University Park, PA 16802
814-863-3411
poultry.cas.psu.edu/

Rhode Island
University of Rhode Island
9 E. Alumni Avenue, Room 137
Kingston, RI 02881
401-874-2970
www.uri.edu/ce/

South Carolina
Clemson University
Animal and Veterinary Sciences
  Department
129 Poole Agricultural Center
Box 340361
Clemson, SC 29634-0361
864-656-3427
cufp.clemson.edu/avs/
virtual.clemson.edu/groups/public

South Dakota
South Dakota State University
Agricultural Hall 154
P.O. Box 2207D
Brookings, SD 57007
605-688-4792

Tennessee
The University of Tennessee
121 Morgan Hall
P.O. Box 1071
Knoxville, TN 37901
865-974-7114
www.utextension.utk.edu

Texas
Texas A&M University
Poultry Science Department
Kleberg Center, Room 101
College Station, TX 77843-2472
979-845-1931
gallus.tamu.edu/departmental.
  html

Utah
Utah State University
4900 Old Main Hill
Logan, UT 84322
435-797-2200
www.ext.usu.edu

Vermont
University of Vermont
College of Agriculture and Life
  Sciences
601 Main Street
Burlington, VT 05405
802-656-2990
ctr.uvm.edu/ext/

Virginia
Virginia Polytechnic Institute
  and State University
Department of Animal and
  Poultry Sciences
101 Hutcheson Hall
Blacksburg, VA 24061-0306
540-231-9185
www.apsc.vt.edu
www.ext.vt.edu

Washington

Washington State University
421 Hulbert Hall
P.O. Box 646230
Pullman, WA 99164
509-335-4561
cahe.wsu.edu

West Virginia

West Virginia University
P.O. Box 6031
Morgantown, WV 26506-6031
304-293-5691
www.wvu.edu\~exten

Wisconsin

University of Wisconsin
Department of Animal Sciences
Animal Science Building
1675 Observatory Drive
Madison, WI 83706-1284
608-263-4300
www.poultry.wisc.edu
www.uwex.edu/ces/cty/grant/index.
   html

Wyoming

University of Wyoming
P.O. Box 3354
Laramie, WY 82071
307-766-5124

ASSOCIATIONS

Associations related to poultry rearing, processing, health and management, and so on are listed as additional sources of information on specific topics and for information about membership. By joining an association, whether local or national, you will be introduced to other like-minded individuals and can help to foster sound poultry management practices.

Alabama Poultry & Egg
  Association
P.O. Box 240
Montgomery, AL 36101
334-265-2732
www.alabamapoultry.org

Alberta Turkey Producers
212 8711A-50 Street
Edmonton, Alberta T6B 1E7
CANADA
780-465-5755
www.abturkey.ab.ca

American Association of Avian
  Pathologists
382 West Street Road
Kennett Square, PA 19348
610-444-4282
www.vm.iastate.edu/aaap

American Association of Meat
  Processors
P.O. Box 269
Elizabethtown, PA 17022
717-367-1168
www.aamp.com

American Farm Bureau Federation
225 Touhy Avenue
Park Ridge, IL 60068
847-685-8600
www.fb.com

American Society of Agricultural
  Engineers
2950 Niles Road
St. Joseph, MI 49085-9659
616-429-0300
www.asae.org

American Veterinary Medical
  Association
1931 N. Meacham Road, Suite 100
Schaumburg, IL 60173
847-925-8070
www.avma.org

Animal Health Institute
1325 G Street NW, Suite 700
Washington, DC 20005
202-637-2440
www.ahi.org

Arkansas Poultry Federation
P.O. Box 1446
Little Rock, AR 72203
501-375-8131

California Poultry Industry
  Federation
3117A McHenry Avenue
Modesto, CA 95350
888-822-4004
www.cpif.org

Delmarva Poultry Industry, Inc.
RD 6, Box 47
Georgetown, DE 19947
302-856-9037
www.dpichicken.org

Florida Poultry Federation
4508 Oak Fair Blvd., Suite 290
Tampa, FL 33610
813-628-4551

Food Distributors International
201 Park Washington Court
Falls Church, VA 22046
703-532-9400
www.fdi.org
www.nawga-ifda.org

Food Processing Machinery &
  Supplies Association
200 Daingerfield Road
Alexandria, VA 22314
800-331-8816
www.fpmsa.org

Food Safety Consortium
110 Agriculture Building
University of Arkansas
Fayetteville, AR 72701
501-575-5647

Georgia Poultry Federation
P.O. Box 763
Gainesville, GA 30503
770-532-0473

Indiana State Poultry Association
Purdue University
1151 Lilly Hall, Room G117
West Lafayette, IN 47906
765-494-8517

Kansas Poultry Association
Kansas State University
Dept. of Animal Sciences
139 Call Hall
Manhattan, KS 66506-1600
785-532-1201
www.oznet.ksu.edu/

Kentucky Poultry Federation
P.O. Box 21829
Lexington, KY 40522
606-266-8375

Louisiana Poultry Federation
214 Knapp Hall
Louisiana State University
Baton Rouge, LA 70803
225-388-2219
www.agctr.lsu.edu

Meat & Poultry Association
of Hawaii
311 Pacific Street
Honolulu, HI 96817
808-585-2900

Midwest Poultry Federation
2380 Wycliff Street
St. Paul, MN 55114
651-646-4553

Mississippi Poultry Association
P.O. Box 13309
Jackson, MS 39236
601-355-0248

Missouri Poultry Federation
225 E. Capitol Avenue
Jefferson City, MO 65101
573-761-5610

National Association of Meat
Processors
1920 Association Drive, Suite 400
Reston, VA 20191-1547
800-368-3043
www.namp.com

National Association of State
Depts. of Agriculture
1015 15th Street NW, Suite 930
Washington, DC 20005
202-296-2622
www.eatchicken.com

National Grocers Association
1825 Samuel Morse Drive
Reston, VA 20190
703-437-5300
www.nationalgrocers.org

National Poultry & Food
Distributors Assoc.
958 McEver Road Ext., Suite B5
Gainesville, GA 30506
770-535-9901
www.npfda.org

National Renderers Association
801 N. Fairfax Street, Suite 207
Alexandria, VA 22314
703-683-0155
www.renderers.org

National Restaurant Association
1200 17th Street NW
Washington, DC 20036
800-424-5156
www.restaurant.org

National Turkey Federation
1225 New York Avenue NW,
Suite 400
Washington, DC 20005
202-898-0100
www.eatturkey.com

Nebraska Poultry Industry
A103 Animal Science
University of Nebraska Lincoln
P.O. Box 830908
Lincoln, NE 68583-0908
402-472-2051

North Carolina Poultry
  Federation
4020 Barrett Drive, Suite 102
Raleigh, NC 27609
919-783-8218

Pacific Egg & Poultry Association
1521 I Street
Sacramento, CA 95814
916-441-0801
www.pacificegg.org

Packaging Machinery
  Manufacturers Institute
4350 N. Fairfax Drive, Suite 600
Arlington, VA 22203
703-243-8555
www.packexpo.com

Poultry Industry Council
RR 2, 483 Arkell Road
Guelph, ON N1H 6H8
519-837-0284
CANADA
www.easynet.ca/~pic

Poultry Science Association
1111 N. Dunlap Avenue
Savoy, IL 61874
217-356-3182
www.psa.uiuc.edu

South Carolina Poultry Federation
1921A Pickens Street
Columbia, SC 29201
803-779-4700

South Dakota Poultry Industries
  Association
Animal and Range Sciences
  Department, SDSU
Box 2170
Brookings, SD 57007
605-688-5409
www.abs.sdstate.edu/ars/index.htm

Texas Poultry Federation
P.O. Box 9589
Austin, TX 78766
512-451-6816

U.S. Poultry & Egg Association
1530 Cooledge Road
Tucker, GA 30084
770-493-9401
www.poultryegg.org

United States Animal Health
  Association
P.O. Box K227
8100 Three Chopt Road
Richmond, VA 23288
804-285-3210
www.usaha.org

Virginia Poultry Federation
P.O. Box 552
Harrisonburg, VA 22801
540-433-2451

Sources of Supplies and Equipment

Many of the supplies and equipment needed for the small poultry flock may be found at the local feed store, hatchery, or other agriculture-supply outlets. Some of the large mail-order houses, such as Sears, also offer agriculture supplies.

The following sources of stock, equipment, and veterinary supplies are listed for your convenience. No endorsement is expressed or implied.

Agri-Equipment International, Inc.
P.O. Box 8401
Greenville, SC 29604
877-550-4709
Bags for dressed poultry.

Anderson Box Company
4030 Vincennes Road
Indianapolis, IN 46268
317-879-4400
*Eggs, cartons, cases, and fillers
Bags for dressed poultry.*

Ashley Machine, Inc.
901 N. Carver Street
P.O. Box 2
Greensburg, IN 47240
812-663-2180
Killing cones, knives, dressing equipment, processing equipment.

Beacon Systems Manufacturing Co.
Route 1, Box 354-A
Buffalo, MO 65622
417-345-2266
Brooders, egg baskets, nests, egg candlers (hand), scales, feeders, waterers.

Big Dutchman
P.O. Box 1017
Holland, MI 49423-1017
616-392-5981
Brooders, nests, feeders, waterers.

Brower Equipment Co.
P.O. Box 2000
Houghton, IA 52631-2000
319-469-4141
Brooders, egg candlers (hand), scales, washers, incubators (small), killing cones, knives, dressing equipment, processing equipment.

Cutler's Supply
3805 Washington Road
Carsonville, MI 48419
810-657-9450
General supplies.

Dussek Campbell, Inc.
National Wax Division
P.O. Box 549
Skokie, IL 60076
847-679-6300
www.dussekwax.com
Wax for defeathering.

First State Packaging, Inc.
511 Naylor Mill Road
Salisbury, MD 21801
410-546-1008
Bags for dressed poultry.

First State Veterinary Supply
P.O. Box 190
Parsonburg, MD 21849
800-950-8387

G.Q.F. Manufacturing Company
2343 Louisville Road
P.O. Box 1552
Savannah, GA 31402-1552
912-236-0651
www.gqfmfg.com
*Wire pens, incubators (small),
brooders, supplies.*

Humidaire Incubator Company
217 W. Wayne Street
P.O. Box 9
New Madison, OH 45346
800-410-6925
www.411web.com (keyword
  "humidaire")
*Redwood cabinet incubators, small
incubators.*

Inman Hatcheries
P.O. Box 616
Aberdeen, SD 57402
800-843-1962
www.inmanhatcheries.com
Chicks of various breeds.

Jeffers Vet Supply
P.O. Box 100
Dothan, AL 36302
800-533-3377
www.jefferspet.com
Medications and general supplies.

Kent Co., Inc.
3030 NE 188th Street
Miami, FL 33180
305-944-4041
*Egg candlers (hand), scales, wash-
ers, incubators (small), processing
equipment, smoke houses.*

Loveland Industries, Inc.
14520 Weld County Road 64
Greeley, CO 80632-1289
970-356-8920
Catching hooks.

Lyon Electric Company
1690 Brandywine Avenue
Chula Vista, CA 91911-6021
619-216-3400
*Incubators, brooders, parts, and
accessories; cannibalism-control
equipment.*

Max-Flex
U.S. Route 219
Lindside, WV 24951
800-356-5458
www.maxflex.com
Electroplastic fencing.

Murry McMurray Hatchery
191 Closz Drive
Webster City, IA 50595
515-832-3280
www.mcmurrayhatchery.com
Chicks, general supplies, books.

NASCO Farm & Ranch
901 Janesville Avenue
Fort Atkinson, WI 53538-0901
800-558-9595
www.enasco.com
Full line of general supplies; egg candlers (hand), scales, washers; killing cones, knives, and other dressing equipment.

National Bag & Tag Company
721 York Street
Newport, KY 41071
Bands; identification, leg and wing.

Northco Industries, Inc.
P.O. Box 718
Luverne, MN 56156-0718
507-283-4411
Brooders.

Omaha Vaccine Company
P.O. Box 7228
Omaha, NE 68107
800-367-4444
www.omahavaccine.com
Medications and supplies.

Peterson Poultry Supplies
P.O. Box 39
Wallburg, NC 27373
336-769-0392
Full line of supplies and books.

Petersyme Incubator Company
P.O. Box 308
Gettysburg, OH 45328-0308
888-255-0067
Incubators (small).

Pickwick Company
1870 McCloud Place NE
Cedar Rapids, IA 52402
319-393-7443
Limited supplies.

Premier Fence Supply
2031 300th Street
Washington, IA 52353
800-282-6631
Electroplastic fencing.

Safeguard Products, Inc.
P.O. Box 8
New Holland, PA 17557
800-433-1819
www.safeguardproducts.com
Wire cages, components, tools, accessories.

Sand Hill Preservation Center
1878 230th Street
Calamus, IA 52729
319-246-2299
Chicks in rare and heirloom breeds.

Shenandoah Manufacturing
  Company, Inc.
1070 Virginia Avenue
Harrisonburg, VA 22802
800-476-7436
Brooders, nests, feeders, waterers.

Smith Poultry & Game Bird
  Supplies
14000 West 215th Street
Bucyrus, KS 66014
913-879-2587
www.poultrysupplies.com
Books and general supplies.

Strecker Supply Company
P.O. Box 190
Parsonburg, MD 21849
800-765-0065
Medications.

Stromberg's Chicks and Gamebirds
  Unlimited
P.O. Box 400
Pine River, MN 56474
800-720-1134
www.strombergschickens.com
*Chicks; full line of supplies and books;
feeders, waterers, incubators (small).*

Waterford Corporation
404 North Link Lane
Fort Collins, CO 80524
800-525-4952
www.waterfordcorp.com
Electroplastic fencing.

GLOSSARY

Abdomen. Area between the keel and pubic bones.

Air cell. Air space usually found in the large end of the egg.

Albumen. The white of the egg.

American breeds. Those breeds developed in America and having common characteristics such as yellow skin, nonfeathered shanks, red earlobes. All lay brown eggs except the Lamonas, which produce white eggs.

Axial feather. The short wing feather between the primaries and secondaries.

Baby chick. Newly hatched chick before it has been fed or watered.

Bantam. Diminutive fowl. Some are distinct breeds, others are miniatures of large breeds.

Beak. Upper and lower mandibles of chickens, turkeys, pheasants, peafowl, etc.

Beak trim. To remove a part of the beak to prevent feather pulling or cannibalism.

Bean. Hard protuberance on the upper mandible of waterfowl.

Bill. The upper and lower mandibles of waterfowl.

Blood spot. Blood in an egg caused by a rupture of small blood vessels, usually at the time of ovulation.

Breast. The forward part of the body between the neck and the keel bone.

Breast blister. Enlarged, discolored area or store in the area of the keel bone.

Breed. A group of fowl related by ancestry and breeding true to certain characteristics such as body shape and size.

Broiler-fryer. Young chickens under 12 weeks of age of either sex that are tender-meated with soft, pliable, smooth-textured skin and flexible breastbone cartilage.

Brooder. Heat source for starting young birds.

Broody. Maternal instinct causing the female to set or want to hatch eggs.

Candle. To determine the interior quality of an egg through the use of a special light in a dark room.

Cannibalism. A vice usually manifested by toe picking, vent picking, or feather pulling.

Capon. A castrated male fowl having undeveloped comb and wattles, and longer hackles, saddle, and tail feathers than the normal male.

Chalazae. White, twisted, ropelike structures that tend to anchor the egg yolk in the center of the egg, by their attachment to the layers of thick albumen.

Chalaziferous layer. Thin layer of thick white surrounding the yolk, continuous with the chalazae.

Cloaca. The common chamber or receptacle for the digestive, urinary, and reproductive systems.

Cock. A male bird over 12 months of age.

Cockerel. A male bird under 12 months of age.

Coccidiostat. A drug used to control or treat coccidiosis disease.

Comb. The fleshy prominence on the top of the head of fowl.

Crop. An enlargement of the gullet where food is stored and prepared for digestion.

Crossbred. The first generation resulting from crossing two different breeds or varieties, or the crossing of first-generation stock with first-generation stocks resulting from the crossing of other breeds and varieties.

Cull. A bird not suitable to be in a laying or breeding pen or not suitable as a market bird.

Culling. The act of removing unsuitable birds from the flock.

Drake. Male duck.

Dub. To trim the comb and wattles close to the head.

Duck. Any member of the family Anatidae and specifically a female.

Duckling. The young of the family Anatidae.

Earlobe. Fleshy patch of skin below the ear. It may be red, white, blue, or purple, depending upon the breed of chicken.

Embryo. A young organism in the early stages of development, as before hatching from the egg.

Face. Skin around and below the eyes.

Flight feathers. Primary feathers of the wing, sometimes used to denote the primaries and secondaries.

Follicle. Thin, highly vascular ovarian tissue containing the growing ovum.

Foot-candle. The density of light striking each and every point on a segment of the inside surface of an imaginary 1-foot-radius sphere with a 1-candle-power source at the center.

Fowl. Term applied collectively to chickens, ducks, geese, etc., or the market class designation for old laying birds.

Gander. Male goose.

Germinal disc, or blastodisc. Site of fertilization on the egg yolk.

Gizzard. Muscular stomach. Its main function is grinding food and partial digestion of proteins.

Goose. The female goose as distinguished from the gander.

Gosling. A young goose of either sex.

Grit. The hard, insoluble materials fed to birds to provide a grinding material in the gizzard.

Gullet, or **esophagus.** The tubular structure leading from the mouth to the glandular stomach.

Hackle. Plumage on the side and rear of the neck of fowl.

Hen. A female fowl more than 12 months of age.

Hock. The joint of the leg between the lower thigh and the shank.

Horn. Term used to describe various color shadings in the beak of some breeds of fowl such as the Rhode Island Red.

Hover. Canopy used on brooder stoves to hold heat near the floor when brooding young stock.

Infundibulum, or **funnel.** Part of the oviduct that receives the ova (egg yolks) when they are ovulated. Fertilization of the egg takes place in the infundibulum.

Isthmus. Part of the oviduct where the shell membranes are added during egg formation.

Keel bone. Breastbone or sternum.

Litter. Soft, absorbent material used to cover floors of poultry houses.

Magnum. Part of the oviduct that secretes the thick albumen or white during the process of egg formation.

Mandible. The upper or lower bony portion of the beak.

Mash. Mixture of finely ground grains.

Meat spots. Generally, blood spots that have changed in color due to chemical action.

Molt. To shed old feathers and grow new ones.

Oil sac, or uropygial gland. Large oil gland on the back at the base of the tail — used to preen or condition the feathers.

Ova. Round bodies (yolks) attached to the ovary. These drop into the oviduct and become the yolk of the egg.

Oviduct. Long, glandular tube where egg formation takes place; leads from the ovary to the cloaca. It is made up of the funnel, magnum, isthmus, uterus, and vagina.

Pendulous crop. Crop that is usually impacted and enlarged and hangs down in an abnormal manner.

Plumage. The feathers making up the outer covering of fowl.

Poult. A young turkey.

Primaries. The long, stiff flight feathers at the outer tip of the wing.

Pubic bones. The thin terminal portion of the hip bones that forms part of the pelvis. Used as an aid in judging productivity of laying birds.

Pullet. Female chicken less than 1 year of age.

Recycle, or force molt. To force into a molt with a cessation of egg production.

Relative humidity. The percentage of moisture saturation of the air.

Replacements. Term used to describe young birds that will replace an old flock.

Roasters. Young chickens of either sex, usually 3 to 5 months of age, that are tender-meated with soft, pliable, smooth-textured skin and with a breastbone cartilage somewhat less flexible than the broiler-fryer.

Roost. A perch on which fowl rest or sleep.

Secondaries. The large wing feathers adjacent to the body, visible when the wing is folded or extended.

Sex-linked. Any inherited factor linked to the sex chromosomes of either parent. A plumage color difference between the male and female progeny of some crosses is an example of sex-linkage. Useful in sexing day-old chicks.

Shell membranes. The two membranes attached to the inner eggshell. They normally separate at the large end of the egg to form air cells.

Slip. A male from which all of both testicles was not removed during the caponizing operation.

Snood. Fleshy appendage on the head of a turkey.

Sperm, or **spermatozoa.** The male reproductive cells capable of fertilizing the ova.

Spur. The stiff, horny process on the legs of some birds. Found on the inner side of the shanks.

Standard-bred. Conforming to the description of a given breed and variety as described in the *American Standard of Perfection*.

Sternum. The breastbone or keel.

Stigma. The suture line or nonvascular area where the follicle ruptures when the mature ovum is dropped.

Strain. Fowl of any breed usually with a given breeder's name and which has been reproduced by closed-flock breeding for five generations or more.

Testicles, or **testes.** The male sex glands.

Trachea, or **windpipe.** That part of the respiratory system that conveys air from the larynx to the bronchi and to the lungs.

Undercolor. Color of the downy part of the plumage.

Uterus. The portion of the oviduct where the thin white, the shell, and shell pigment are added during egg formation.

Vagina. Section of the oviduct that holds the formed egg until it is laid.

Vent, or **anus.** The external opening from the cloaca.

Vitelline membrane. The thin membrane that encloses the ovum.

Wattles. The thin pendant appendages at either side of the base of the beak and upper throat, usually much larger in males than in females.

Windpuffs. Air trapped under the outer skin as a result of rupturing the air sacs during caponization.

Xanthophyll. One of the yellow pigments found in green plants, yellow corn, fatty tissues, and egg yolks.

Yolk. Ovum, the yellow portion of the egg.

INDEX

Note: Page numbers in *italics* refer to illustrations; those in **boldface** refer to charts.

OTHER STOREY TITLES YOU WILL ENJOY

Basic Butchering of Livestock & Game by John J. Mettler Jr., DVM. Provides, clear, concise, step-by-step instructions for individuals interested in slaughtering their own meat. 208 pages. Paperback. ISBN 0-88266-391-7.

Building Small Barns, Sheds & Shelters by Monte Burch. Covers tools, materials, foundations, framing, sheathing, wiring, plumbing, and finish work for barns, woodsheds, garages, fencing, and animal housing. 248 pages. Paperback. ISBN 0-88266-245-7.

The Chicken Health Handbook by Gail Damerow. Packed with relevant information for all breeds and ages, this book covers nutrition, disease, immunity, and anatomy. Written for the chicken fancier and nonspecialist. 352 pages. Paperback. ISBN 0-88266-611-8.

Fences for Pasture & Garden by Gail Damerow. The complete guide to choosing, planning, and building today's best fences: wire, rail, electric, high-tension, temporary, woven, and snow. 160 pages. Paperback. ISBN 0-88266-753-X.

How to Build Small Barns & Outbuildings by Monte Burch. This book takes the mystery out of small-scale construction. Projects are offered with complete plans and instructions. 288 pages. Paperback. ISBN 0-88266-773-4.

Storey's Guide to Raising Turkeys by Leonard S. Mercia. Offers current information on selecting the right turkey, feeding and housing, breeding and management, flock health, and processing. 208 pages. Paperback. 1-58017-261-X.

Wild Turkeys: Hunting and Watching by John J. Mettler Jr., DVM. This book explains how to choose clothing, firearms, and equipment; addresses turkey distribution, feeding patterns, and mating rituals; how to dress a turkey carcass, mount a prize bird, and photograph turkeys in the wild. 176 pages. Paperback. ISBN 1-58017-069-2.

These books and other Storey books are available at
your bookstore, farm store, garden center, or directly from
Storey Books, 210 MASS MoCA Way, North Adams, MA 01247,
or by calling 1-800-441-5700.
Or visit our Web site at www.storey.com.